NUMER... IN ...

For
[Pre University and Higher Secondary Examinations of Punjab, Haryana, Himachal Pradesh, Chandigarh, Jammu and Kashmir]

By
B. S. BAHL
Formerly Principal, D.A.V. College, Jalandhar.
Member Senate, Guru Nanak Dev University,
Amritsar

AND

G. D. SHARMA
Vice Principal and Head, Post-Graduate Department of Chemistry,
D.A.V. College, Jalandhar ; Member Board of Studies
in Chemistry, Member Science Faculty, Guru Nanak Dev
University, Amritsar

AND

ARUN BAHL
M.S. (Boston). Ph.D.(Edinburgh), C. Chem. M.R.S.C. (London);
Lecturer in Chemistry;
Panjab University · Chandigarh

NEW
SIXTH EDITION

S. CHAND
AN ISO 9001: 2000 COMPANY

S. CHAND & COMPANY LTD.
RAM NAGAR, NEW DELHI-110 055

S. CHAND & COMPANY LTD

(An ISO 9001 : 2000 Company)

Head Office : 7361, RAM NAGAR, NEW DELHI - 110 055
Phones : 23672080-81-82, 9899107446, 9911310888;
Fax : 91-11-23677446
Shop at: **schandgroup.com** E-mail: **schand@vsnl.com**

Branches:

- 1st Floor, Heritage, Near Gujarat Vidhyapeeth, Ashram Road, **Ahmedabad**-380 014. Ph. 27541965, 27542369.
- No. 6, Ahuja Chambers, 1st Cross, Kumara Krupa Road, **Bangalore**-560 001. Ph : 22268048, 22354008
- 238-A M.P. Nagar, Zone 1, **Bhopal** - 462 011. Ph : 4274723,
- 152, Anna Salai, **Chennai**-600 002. Ph : 28460026
- S.C.O. 2419-20, First Floor, Sector- 22-C (Near Aroma Hotel), **Chandigarh**-160022, Ph-2725443, 2725446
- 1st Floor, Bhartia Tower, Badambadi, **Cuttack**-753 009, Ph-2332580; 2332581,
- 1st Floor, 52-A, Rajpur Road, **Dehradun**-248 001. Ph : 2740889, 2740861,
- Pan Bazar, **Guwahati**-781 001. Ph : 2738811
- Sultan Bazar, **Hyderabad**-500 195. Ph : 24651135, 24744815,
- Mai Hiran Gate, **Jalandhar** - 144008 . Ph. 2401630, 5000630
- A-14 Janta Store Shopping Complex, University Marg, Bapu Nagar, **Jaipur** - 302 015, Phone : 2719126
- 613-7, M.G. Road, Ernakulam, **Kochi**-682 035. Ph : 2378207
- 285/J, Bipin Bihari Ganguli Street, **Kolkata**-700 012. Ph : 22367459, 22373914
- Mahabeer Market, 25 Gwynne Road, Aminabad, **Lucknow**-226 018. Ph : 2626801, 22
- Blackie House, 103/5, Walchand Hirachand Marg , Opp. G.P.O., **Mumbai**-400 001, Ph : 22690881, 22610885
- Karnal Bag, Model Mill Chowk, Umrer Road, **Nagpur**-440 032 Ph : 2723901, 2777666, 2720523
- 104, Citicentre Ashok, Govind Mitra Road, **Patna**-800 004. Ph : 2300489, 2302100,
- 291/1, Ganesh Gayatri Complex, 1st Floor, Somwarpeth, Near Jain Mandir, **Pune**-411011. Ph : 64017298
- Flat No. 104, Sri Draupadi Smriti Apartment, East of Jaipal Singh Stadium, Neel Ratan Street, Upper Bazar, **Ranchi**-834001. Ph: 2208761
- Kailash Residency, Plot No. 4B, Bottle House Road, Shankar Nagar, **Raipur**. Ph. 099812

© *Copyright Reserved*

All rights reserved. No part of this publication may be reproduced, stored in a retrieval system or transmitted, in any form or by any means, electronic, mecha photocopying, recording or otherwise, without the prior permission of the Publ

First Edition 1958
Subsequent Editions and Reprints 1961, 65, 67, 69, 72, 79, 83, 85, 89, 90, 94, 96, 97, 2000, 2001, 2002, 2003, 2005, 2006
Reprints 2008 (Twice)
ISBN : 81-219-0858-2
Code : 04 016

PRINTED IN INDIA
By Rajendra Ravindra Printers (Pvt.) Ltd., Ram Nagar,
New Delhi-110 055 and published by S. Chand & Company Ltd.,
7361, Ram Nagar, New Delhi-110 055

PREFACE TO SIXTH EDITION

This book is designed for Pre-University and Higher Secondary students. This new edition retains the objectives of the first edition : (1) to *provide* students with simple and direct methods for solving numerical problems; and (2) to provide teachers with supplementary problems and exercises for use in tutorials and as homework assignment.

This new edition is designed to be **self-teaching**. Each chapter contains :

(1) **Necessary Principles** for solving a given numerical problem. Explanations are given in plain and simple language.

(2) **Examples** using principles discussed, **PLUS**, detailed step by step solutions. We have tried to make problem-solving easier for students.

(3) **End-of-Chapter Problems** that enable students to test their knowledge.

But that's not all ! This new edition offers an abundance of Latest Solved University Questions.

June, 1983

B. S. Bahl
G. D. Sharma
Arun Bahl

Problem solving is not so disappointing if you have enough practice

GENERAL ADVICE ON THE SOLVING OF CHEMICAL PROBLEMS

1. **Read the Problem Carefully.** Note exactly what is given and what is sought. Be sure that you understand all the Chemical Principles that are involved.

2. **Plan, in Detail, Just how the Problem is to be Solved.** Each step in a calculation should be clearly set out. It is better to give too much explanation than too little.

3. **Always State the Units of Chemical quantities.** Failure to do this may lead to confusion.

4. **Always Check Your Answer** by working out an approximate answer. This will prevent the giving of an impossible solution to a problem, such as 1 g of hydrogen combines with 800 g of oxygen.

5. **Write Clearly.** Particularly the numbers, and do not over-crowd the work. Many errors are made through misreading an earlier part of the calculation.

CONTENTS

Chapter		Pages
1. Units of Measurements	...	1—7
2. Laws of Chemical Combination	...	8—19
3. Percentage Composition	...	20—29
4. Mole Concept	...	30—42
5. Empirical and Molecular Formulae	...	43—57
6. Gas Laws	...	52—69
7. Diffusion of Gases	...	70—81
8. Molecular Weights	...	82—91
9. Equivalent Weights	...	92—121
10. Atomic Weights	...	122—137
11. Oxidation and Reduction	...	138—150
12. Volumetric Analysis	...	151—185
13. Problems Based on Equations	...	186—204
14. Problems Based on Equations (Contd)	...	205—218
15. Problems Based on Equations (Contd)	...	219—228
16. Measurement and Uncertainty	...	229—235
17. Glossary	...	236—244
Appendices	...	245—260
Log Tables	...	261—269

Units of Measurement

In the scientific world a uniform system of expressing weights and measures has been established. Most of the calculation work in chemisty is concerned with measurements of mass, volume, temperature and pressure. It is necessary that the students of chemistry should have a working knowledge of all the fundamental units of measurements.

METRIC SYSTEM

The metric system is used in all branches of science and is the legal system of measurements in almost all the countries of the world except in Great Britain and United States. It is often called *centimetre—gram—second* or **CGS system**. *It is a decimal system and derives its name metric from the basic unit of length, the 'metre'.*

Length. The fundamental unit of length is metre (m). It is divided into 100 equal parts, called centimetres (cm). A centimetre is further divided into 10 millimetres (mm). Thus,

$$10 \text{ millimetres} = 1 \text{ centimetre}$$
$$10 \text{ centimetres} = 1 \text{ decimetre}$$
$$10 \text{ decimetres} = 1 \text{ metre}$$
$$10 \text{ metres} = 1 \text{ decametre}$$
$$10 \text{ decametres} = 1 \text{ hectometre}$$
$$10 \text{ hectometres} = 1 \text{ kilometre}$$

Volume. The unit of volume in the *CGS* system is the **Cubic Centimetre** (cc or cm^3). B still larger unit of volume is litre (l). Thus,

1000 millilitres (ml) = 1 litre (l)

1 litre (l) = 1000·27 cc

1 millilitre = 1·000027 cc

Difference between a Millilitre and a Cubic centimetre. For practical purposes a cubic centimetre (cc) is considered same as a millilitre. But there is a slight difference between the two.

At one time it was thought that the mass of 1 kilogram of water should be equal to the mass of 1000 cc of water at 3·98°C (the temperature at which water is most dense). Precise measurements have now shown that 1 kilogram of water at this temperature measures actually 1000·27 cc instead of 1000 cc. In modren practice, litre is considered to be the unit of volume on the *CGS* system and 1 kilogram of water at 3·98°C occupies 1 litre or 1000 millilitres. But 1 kilogram of water also occupies 1000·027 cc at this temperature. This means that 1000 millilitres are equal 1000·027 cc at 3·98°C. Thus,

$$1000 \text{ ml} = 1000·027 \text{ cc}$$

Or $1 \text{ ml} = 1·000027 \text{ cc}$

This difference, of course, has no practical significance in elementary chemical problems.

Mass. Mass is the measure of the quantity of matter.

Weight. It is the pull of gravity on any mass and is used as the measure of mass.

The *CGS* unit of mass is the **Gram** (g) which is the weight of 1 ml of water at 3·98°C. A larger unit of mass the **kilogram** (kg) *is equal to 1000 grams* A smaller unit of mass is *milligram* (mg). Thus,

1 kilogram (kg) = 1000 grams (g)

1 gram (g) = 1000 milligrams (mg)

THE BRITISH OR THE U.S. SYSTEM

It is the official system in these two countries. This system is often called FPS system. The **FPS System** means *Foot, Pound, Second* System. The unit of length is *foot*, the unit of mass is the *pound* and the unit of time is the *second*. The unit of volume will thus be **Cubic Foot** (*foot* × *foot* × *foot*).

Conversion Table for Length

1 inch (in) = 2·54 cm = 25·4 mm
1 foot (ft) = 30·48 cm = 0·3048 m

Conversion Table for Mass or Weight

1 ounce (oz) = 28·35 g
1 pound (lb) = 453·6 g
1 kilogram   = 2·2046 (lb)
1 mg   = 0·01543 grain
1 grain   = 15·43 grains

Conversion Table for Capacity

1 fluid dram   = 3·70 ml
1 fluid ounce (fl. oz) = 29·57 ml
1 cubic inch   = 16·39 ml
1 cubic foot   = 28·32 litres
1 litre   = 1·057 quarts
1 quart   = 0·9463 litres
1 gallon   = 3·785 litres.

MEASUREMENT OF TEMPERATRE

To measure the intensity of heat content of a body or a system, temperature scales have been devised. The most commonly employed scales of temperature are the *Fahrenheit* and *Centigrade*. The former is used for the clinical thermometers or in engineeing while the latter is used in scientific laboratories.

On the **Fahrenheit Scale** (°F) the freezing point of water is defined as 32° and the boiling point as 212° at one atmosphere. The distance between these two fixed levels is divided into 180 equal parts. Each division on this scale corresponds to 1°F.

On the **Centigrade Scale**, the freezing point of water is defined as 0° and the boiling point as 100° at one atmosphere. The distance between these two fixed levels is divided into 100 equal parts. Each division on this scale corresponds to 1°C. *The centigrade scale is also known as* **Celsius Scale**.

Comparison of Two Scales. Between the freezing point and boiling point of water there are 100 centigrade degrees and 180 Fahrenheit degrees.

Thus 100° centigrade = 180° Fahrenheit

or $1°C = \frac{180}{100}$ or $\frac{9}{5}$ °F

or $1°F = \frac{5}{9}$ °C.

For the **interconversion of temperatures on the two scales**, the following formula may be employed :

$$0°C = \frac{5}{9} (0°F - 32)$$

and

$$0°F = \frac{9}{5} (0°C + 32)$$

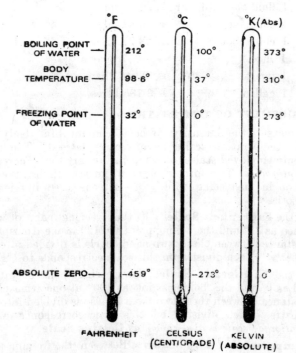

Fig. 1·1. A comparison of the Fahrenheit, Centigrade, and Absolute scales of temperature.

Units of Measurement

Absolute Temperature Scale. It is a new scale for measuring temperature. Like the centigrade scale, it was only 100 degrees with the difference that the zero on this scale is 273° below the zero of the centigrade scale. The scale of temperature is known as *'absolute scale'* and the temperatures reckoned on this scale are called *absolute temperatures*. The zero of this scale *i.e.*, the starting point on this scale is called **Absolute Zero.** There are two methods employed to express temperatures on this scale (a) the **Kelvin scale** (°K) ; and (b) **Rankine Scale** (°R). The Kelvin scale is the **Absolute Centigrade Scale** in which the degrees are expressed in the same way as the centigrade. Thus,

$$0°K = -273°C$$
or
$$273°K = 0°C$$
and
$$373°K = 100°C$$

To convert a temperature from centigrade to Kelvin scale we only add 273 algebraically to the given centigrade temperature. That is,

$$t°C = (t°C + 273)°K$$
or
$$°K = °C + 273$$

Thus 0°C (the freezing point of water)
$$= 0 + 273°K = 273°K$$
$$3°C = (3 + 273)°K \text{ i.e., } 276°K$$
$$50°C = (50 + 273)°K \text{ i.e., } 323°K$$

100°C (boiling point of water) $= (100+273)°K$ *i.e.,* $+373°K$
$$-10°C = (-10 + 273)°K \text{ i.e., } +263°K$$
$$-100°C = (-100 + 273)°K \text{ i.e., } +173°K.$$

The **Rankine Scale** is the **Absolute Fahrenheit Scale.**

Thus, $0°R = -460°F$

i.e., the zero on this scale is 460° below the zero on the Fahrenheit scale.

So that $0°R = -460°F$
or $0°F = 460°R.$

To convert a temperature from Fahrenheit- scale to the Rankine scale, we add 460 algebraically to given Fahrenheit temperature.

Thus, $t°F = (t°F + 460)°R$

or $°R = °F + 460$

Thus, $61°F = (61 + 460)°R$ or $521°F$

Similarly, $100°F = (100 + 460)°R$ or $560°R$

or $32°F$ (the freezing point of water) $= (32+460)°R$
$= 492°R$

and $212°F$ (the boling point of water) $= (212+460)°R$
$= 672°R$

Out of the Kelvin and the Rankine scales, *the former is frequently employed in scientific research.*

Example 1. *Convert $-22°C$ and $10°C$ to Kelvin scale.*

Kelvin scale $(°K)$ = Centigrade temperature + 273

$-22°C = -22 + 273 = 251°K$

$10°C = 10 + 273 = 283°K$

Example 2. *Convert $120°K$ and $468°K$ to Centigrade scale.*

Centigrade temperature $(°C)$ = Kelvin temperature -273

Thus $120°K = 120 - 273 = -153°C$

and $468°K = 468 - 272 = 195°C$

MEASUREMENT OF PREESURE

Pressure *is defined as the force acting on a unit area of the surface.*

Mathematically,

$$\text{Pressure} = \frac{\text{Force acting perpendicular to an area}}{\text{Total area over which force is distributed}}$$

Since air has weight, it exerts a pressure. When we say **'atmospheric pressure'**, we mean the pressure of the overlying air present in the atmosphere.

The pressure exerted by a column of liquid is mathematically equal to the product of the height of the liquid column and its density.

In the **CGS System**, pressure is measured in gram per square centimetre (g/cm^2).

Thus, pressure in g/cm^2 = Height of the column (cm)
  $\times$ density of the liquid (g/cm^3)

Units of Measurement

A column of mercury (density = 13·596 g/cm³) 76 cm (or 760 mm) in height exerts a pressure equal to 1 atmosphere (atm)

or 1 atm = 76 cm of mercury
  = 760 mm of mercury *(at sea level)*.

English system. In the English system pressure is measured in lbs/in² (pounds per square inch). On this scale,

1 atmosphere pressure = 14·7 lb/in² *(at sea level)*.

Thus,

1 atm = 76 cm of Hg = 760 mm of Hg = 29·22 in Hg
  = 14·7 lb per square inch (14·7 psi)
  = 2,116·8 lbs/square foot.

STANDARD TEMPERATURE AND PRESSURE

Standard temperature and pressure (**S.T.P.**) is also referred to as *Normal temperature and pressure* (**N.T.P.**). S.T.P. or N.T.P. denotes a temperature of 0°C or 273°K and pressure of 76 cm or 760 mm of Hg (or 1 atmosphere). When we mention volumes or density of gases, we refer these volumes at N.T.P. or S.T.P. *Actually all standard measurements in respect of gases are done by reducing their volumes to standard conditions* (S.T.P. or N.T.P.) *for purposes of comparison.*

END-OF-CHAPTER PROBLEMS

1. Convert the following temperatures to their corresponding values on the Kelvin scale :

68°C, —10°C, 150°C, —273°C, 100°C, 200°C.
[**Ans.** 341°K, 263°K, 423°K, 0°K, 373°K, 473°K]

2. Convert 150°K, 760°K, 90°K, 250°K to the degrees on the Centigrade scale. [**Ans.** —123°C, 457°C, 183°C, —230°C]

3. Which of the two will be hotter, water at —88°C, or water at 160°K ? [**Ans.** Water at —88°C]

4. Which of the two will be cooler, water at 50°C, or water at 373°K ? [**Ans.** Water at 50°C]

5. Which of the two pressures is greater

 (a) 76 cm of mercury or 40 inches of mercury ?

 (b) 2 atmospheres or 140 cm of mercury ?

 (c) 740 mm of mercury or 0·4 atmosphere ?
[**Ans.** (a) = 40 inches of mercury ;
  (b) = 2 atmospheres ;
  (c) = 740 mm of Hg]

2

Laws of Chemical Combination

When we study chemical reactions quantitatively, *i.e.*, from the point of view the quantities of the reactants and the products taking part in the reaction, they are found to obey certain laws which are commonly known as the *Laws of Chemical Combination*. These laws have been experimentally verified. These laws are :

(1) *Law of Conservation of Mass.*

(2) *Law of Constant Composition or Definite Proportions.*

(3) *Law of Multiple Proportions.*

(4) *Law of Reciprocal Proportions.*

Laws of Conservation of Mass. This law is also known as the *Law of Indestructibility of Matter*. It states :

"*The total mass of the reacting substances remains the same throughout the change*"

It means that during all changes whether physical or chemical, the total mass of the substances before the change is equal to the total mass of the substances after the change. With the help of this law, it is possible to write equations for various chemical reactions. Most of the calculations based on chemical equations are an application of this law.

Example 1. *When 0.28 g of iron filings were heated in a current of dry air, gave 0.36 g of iron oxide. Find the weight of* oxygen that combines with this weight of metal with the help of the Law of Conservation of Mass.

From the law of conservation of mass we have wt. of oxygen and wt. of iron = wt. of iron oxide

or   Wt. of oxygen + 0·28 = 0·36

Wt. of oxygen which combines with 0·28 g iron

$$= 0\cdot36 - 0\cdot28$$
$$= 0\cdot08 \text{ g.}$$

Law of Constant Composition or Definite Proportions. The law states :

"A chemical compound always consists of the same elements combined together in the same fixed ratio by weight."

Thus calcium oxide may be prepared in many different ways *e.g.*, by the direct union of calcium and oxygen, heating calcium carbonate etc. By whatever method it is prepared, the compound calcium oxide will contain the same elements calcium and oxygen, and in the same ratio by weight *i.e.*, 40 parts by weight of calcium and 16 parts by weight of oxygen. If any of the constituents are present in excess of the required ratio for the formation of a compound, the excess will be left uncombined. Not only the **ratio by wt. of calcium** is constant, but ratios

$$\frac{\text{wt. of calcium}}{\text{wt. of calcium oxide}} \text{ or } \frac{\text{wt. of oxygen}}{\text{wt. of calcium oxide}}$$

will also be constant.

Example 2. *In two experiments 0·259 gram and 0·207 gram of lead were converted into lead chloride, yielding 0·347 gram and 0·278 gram of chloride respectively. Show that the data illustrates the law of constant composition.*

First experiment. Wt. of lead chloride = 0·347 g

Wt. of lead metal = 0·259 g

∴ wt. of chloride which combines with 0·259 g lead

$$= 0\cdot347 - 0\cdot259 = 0\cdot088 \text{ g}$$

Now 0·088 gram chlorine combines with lead = 0·259

1 ,, ,, ,, ,, ,, ,, $= \dfrac{0\cdot259}{0\cdot088}$

$$= 2\cdot92$$

The ratio of chlorine : lead = 1 : 2·92

Second experiment

Wt. of lead chloride = 0·278 g
Wt. of lead metal   = 0·207 g

∴ Wt. of chlorine that combines with 0·207 gram lead

$$= 0.278 - 0.207$$
$$= 0.071 \text{ g}$$

Now 0·071 g chlorine combines with metal

$$= 0.207$$

1 ,, ,, $= \dfrac{0.207}{0.071} = 2.91$

The ratio of chlorine : lead = 1 : 2·91

The two ratios are same. Hence the data illustrates the law of constant composition.

Law of Multiple Proportions. The law states :

"*When same two elements combine to form two or more than two compounds, the weigth of one of these elements which combines with the fixed weight of the other bear a simple ratio to one another*".

To illustrate this law we take the compounds of sulphur and oxygen. These two elements combine to form a number of oxides, SO, S_2O_3, SO_2, SO_3, and SO_4. When the weight of sulphur is fixed in these compounds, we find that weights of oxygen in all these compounds are in the simple whole number ratio. Thus :

	Wt. of sulphur	Wt. of oxgyen	Wt. of S fixed	Wt. of O_2	Ratio of the wt. of Oxygen
SO	32	16	32	16	2
S_2O_3	64	48	32	24	3
SO_2	32	32	32	32	4
SO_3	32	48	32	48	6
SO_4	32	64	32	64	8

Laws of Chemical Combination

The ratio of weights of oxygen in all these oxides comes to be 2 : 3 : 4 : 6 : 8 which is simple and whole number.

Example 3. *Copper combines with oxygen to form two oxides which have the following composition :*

(i) *0·716 g of cuprous oxides contains 0·630 g of copper.*

(ii) *0·398 g of cupric oxide contains 3·318 g of copper.*

Prove that the above data illustrates the law of multiple proportions.

(a) In cupric oxide :

 Wt. of cupric oxide = 0·398 g
 Wt. of copper   = 0·318 g
∴ Wt. of oxygen   = 0·080 g

(b) In cuprous oxide :—

 Wt. of cuprous oxide = 0·716 g
 Wt. of copper   = 0·636 g
∴ Wt. of oxygen   = 0·080 g

In cuprous oxide, wt. of Cu = 0·636 g, wt. of oxygen
                = 0·080 g
In cupric oxide, wt. of Cu = 0·318 g, wt. of oxygen
                = 0·080 g

Since the weights of oxygen are same in both the cases, the weights of copper are in the ratio of 0·636 : 0·318 (*i.e.*, 2 : 1).

Thus we find that when the weights of oxygen are fixed, the wts. of copper are in simple whole number ratio (2 : 1).

This illustrates the law of multiple proportions.

Law of Reciprocal Proportions. The law states :

"When two different elements combine with the same quantity of the third element, the ratio in which they do so will be same or the simple multiple of the proportion in which they unite with each other."

The following examples illustrate the Law of Reciprocal Proportions.

Example 4. *One gram of hydrogen combines with 15.88 g of sulphur. One gram of hydrogen combines with 7.92 g of oxygen, one gram of sulphur combines with 0.998 g of oxygen. Show that these data illustrate the Law of Reciprocal Proportions.*

It is given that,

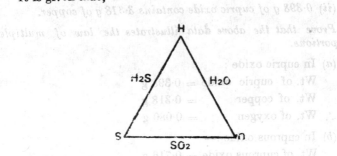

One gram of hydrogen combines with sulphur = 15.88 g

One gram of hydrogen combines with oxygen = 7.92 g

The ratio of the weights of sulphur and oxygen which combines with one gram of hydrogen = 15.88 : 7.92
= 2 : 1 (approx.)

Also, one gram of sulphur combines with oxygen = 0.988 g

The ratio of the weights of sulphur and oxygen
= 1 : 1 (approx)

Thus we find that the ratio of the weights of sulphur and oxygen which combine with same weight of sulphur is a simple multiple of the ratio of the weights of sulphur and oxygen in which they unite with each other.

Example 5. *The percentage composition of nitrous oxide is 63.65 per cent nitrogen and 36.35 per cent oxygen. The percentage composition of water is 11.21 per cent hydrogen and 88.79 per cent oxygen. The percentage composition of ammonia is 82.22 per cent nitrogen and 17.78 per cent hydrogen. Show how these data illustrate the law of reciprocal proportions,*

Laws of Chemical Combination

It is given that:

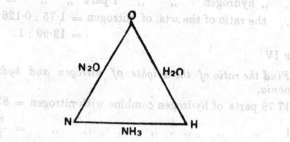

the percentage of N in nitrous oxide = 63·65
" " O " " = 36·35
the percentage of hydrogen in water. = 11·21
" " oxygen " " = 88·79
the percentage of nitrogen in ammonia = 82·22
" " hydrogen " " = 17·78

STEP I

Find the wt. of nitrogen which combines with 1 part by weight of oxygen.

36·35 parts of oxygen combine with nitrogen
= 63·65

∴ 1 " " " " = $\frac{63·65}{36·35}$

= 1·75 parts

STEP II

Find the wt. of hydrogen which combines with 1 part of oxygen.

88·79 parts by weight of oxygen combine with H = 11·21

1 " " " " = $\frac{11·21}{88·79}$ = 0·126

STEP III

Find the ratio of the weights of nitrogen and hydrogen which combine with 1 part of oxygen.

Wt. of nitrogen combining with 1 part of oxygen = 1·75
" " hydrogen " " 1 part " " = 0·126
∴ the ratio of the wts. of nitrogen = 1·75 : 0·126
= 13·99 : 1.

STEP IV

Find the ratio of the weights of nitrogen and hydrogen in ammonia.

17·78 parts of hydrogen combine with nitrogen = 82·22

$$1 \text{ " " " " " " " } = \frac{82 \cdot 22}{17 \cdot 78}$$

$$= 4 \cdot 63.$$

Thus, the ratio of the wts. of nitrogen and hydrogen
= 4·63 : 1.

Now the ratio of the wts. of nitrogen and hydrogen which combine with same weight of oxygen is 13·90 : 1 and the ratio of the wts. of N and H when they combine with each other is 4·63 : 1. We find that ratio of hydrogen being same (one) in both, that of nitrogen is the first case in three times (a simple multiple) that in the second case (4·63 × 3 = 13·89)

This illustrates the Law of Reciprocal Proportions.

END-OF-CHAPTER PROBLEMS

Law of Constant Composition

1. Three different samples of sodium chloride were found to contain sodium and chlorine only and the composition of each sample was

(*i*) 1·6 gm of sodium chloride contained 0·970 g of chlorine and 0·77 g of sodium.

(*ii*) 11·6 g of sodium chloride contained 6·432 g of chlorine and 4·168 g of sodium.

(*iii*) 3·33 g of sodium chloride contained 2·022 g of chlorine and 1·308 g of sodium.

Show that these figures illustrate the Law of Constant Composition.

2. 1·375 g of cupric oxide were reduced by heating in a current of hydrogen and the weight of copper that remained was 1·098 g. In another experiment 1·179 g of copper was dissolved in nitric acid and the resulting copper nitrate converted into cupric oxide by ignition. The weight of cupric oxide formed was 1·476 g. Show that these results illustrate the Law of Definite Proportions.

Laws of Chemical Combination

3. In two experiments 0·812 and 0·7204 g of pure carbon gave 2·934 and 2·638 g of CO_2 respectively when burnt in air. Show that the data illustrates the Law of Constant Composition.

4. 1·4 g of iron combines directly with 0·8 g of sulphur to yield iron sulphide. In another experiment 2·1 g of iron on treatment with H_2SO_4 and sodium sulphide gave 3·3 g of iron sulphide. Show that the above data illustrates the Law of Constant Composition.

5. Weight of copper oxide obtained by treating 2·16 g of metallic copper with nitric acid and subsequent ignition was 2·70 g. In another experiment 1·15 g of CuO on reduction yielded 0·92 g of copper. Show that the results illustrate the law of definite proportions.

6. Zinc sulphate crystals contain 22·65% of zinc and 43·9% of water. If the law of constant composition is true, how much zinc is required to produce 27·4 g of zinc sulphate crystals and what amount of water will they have. [**Ans.** 6·206 g $ZnSO_4$; 12·028 g water

7. The percentage composition of ferrous sulphide is

$$Fe = 63.54\%$$
$$S = 36.46\%$$

If the law of constant composition is true, how much iron sulphide will be obtained from 4 g of iron when heated with 5 g of sulphur ? How much sulphur will be left unconsumed ?
[**Ans.** 6·296 g of FeS; 0·408 g of Sulphur]

8. In an experiment 0·54 g of silver metal on treatment with dil. HNO_3 and subsequent treatment with HCl gave 0·717 g of AgCl. In another experiment 2·64 g of AgCl on analysis gave 2 g of silver metal and 0·64 g of chlorine. Show that the above data illustrates the Law of Constant Composition.

9. A student prepared silver chloride in two of ways with the following results :

Wt. of Ag	Wt. of AgCl
(i) 107·880 g	143·340
(ii) 35·960 g	47·780

Show that these results are in agreement with the Law of Definite Proportions.

10. (i) 2·4 g of magnesium when heated in air formed 4·0 g of magnesium oxide.

(ii) Magnesium oxide obtained by heating magnesium carbonate was found to contain 60% magnesium and 40% oxygen.

Show that the results in (i) and (ii) are in accordance with the law of constant composition.

11. What law do the following data illustrate ?

	Cu	Cl
Cuprous chloride	64·3%	35·7%
Cupric chloride	47·4%	52·6%

(*Venkateswara Pre-Univ.*, 1979)

Law of Multiple Proportions

1. The three oxides of phosphorus contain 43.668, 49.212 and 55.365 per cent of phosphorus.
Show that these figures illustrate the law of multiple proportions.

2. 11 g of an oxide of nitrogen gave 5.6 litres of nitrogen at N.T.P. 15 g of another oxide of nitrogen yielded same volume of nitrogen at N.T.P. Show that these results illustrate the Law of Multiple Proportions.

3. Sulphur trioxide contains 59.45 per cent oxygen. Potassium sulphate contains 18.39 per cent of sulphur, 36.71 per cent of oxygen. Show that in these two compounds sulphur and oxygen illustrate the law of multiple proportions.

4. Sulphur dioxide contains 50.03% sulphur while sulphur trioxide contains 40.03 per cent of sulphur. Show how these compounds illustrate the law of multiple proportions.

5. Methane (CH_4) contains 75% carbon, ethylene (C_2H_4) contains 85.7% carbon, acetylene (C_2H_2) contains 92.22% carbon. Show that the above data illustrates the law of multiple proportions.

6. Three oxides of lead weighing 2.173 g, 1.949 g and 2.316 g respectively gave 2.017 g, 1.688 g and 2.10 g of lead respectively. Show that these figures illustrate the law of multiple proportions.

7. A certain metal forms two oxides containing 44.44 and 60.00 per cent oxygen respectively. Show that these figures illustrate the law of multiple proportions.

8. Copper gives two oxides, on heating one gram of each in hydrogen we get 0.888 g and 0.798 g of metal respectively. Show that these results illustrate the law of multiple proportions.

9. A metal forms two chlorides containing respectively 65.6% and 55.9% chlorine. Show that these figures are in accordance with fundamental laws of chemistry and state the law.

10. State and explain the law of multiple proportions. Two oxides of hydrogen H_2O and H_2O_2, contain 88.8% and 94.07% of oxygen respectively. Show how this data illustrates the law of multiple proportions.

11. A metal is found to give two oxides. On heating one gram of each in a current of hydrogen 0.798 and 0.888 gm of the metal was obtained. Show that the results are in accordance with the law of multiple proportions.

12. On analysis it was found that Litharge, Red oxide and Peroxide of lead contain 92.80%, 90.6% and 86.6% of the metal. Establish the law of multiple proportions from this data.
(Panjab Pre-Univ. 1973)

13. Two chlorides of a metal contain 37.56 and 54.20 per cent of chlorine. Show that these figures illustrate the law of multiple proportions.
(Kurukshetra Pre-Univ., 1973)

Laws of Chemical Combination

14. A metal forms two oxides. One of the oxides has 20% oxygen and the other 11·1% oxygen. Show which of the laws of chemical combination is illustrated here.

(Sir Venkateswara Pre-Univ. 1973)
[**Ans**. Law of multiple proportions]

15. On heating oxides of lead in a current of hydrogen, the following results were obtained. Show that they are in agreement with the law of multiple proportions.

(i) 1·393 gm of litharge gave 1·293 gm of lead.
(ii) 2·173 gm of lead dioxide gave 1·882 gm of lead.
(iii) 1·712 gm of red lead gave 1·552 gm of lead.

(Guru Nanak Dev. Pre Univ., 1977)

16. Elements A and B combine to form three different compounds : 0·3 g of A + 0·4 g of B → 0·7 g of compound x ; 18 g of A + 48 g of B → 66 g of compound Y ; 40 g of A + 159·99 g of B → 199·9 of Z. Show that the *Law of Multiple Proportions* is verified by the data given above.

(Kurukshetra Pre-Univ., 1977)

17. Phosphorus is found to form three oxides containing 43·67, 49·29 and 56·36% of phosphorus respectively. Show that these figures illustrate the Law of Multiple Proportions.

(Punjabi Pre Univ., 1978)

18. Ferrous chloride and Ferric chloride contain 44·1% and 34·43% Iron. Are these figures in accordance with the Law of Multiple Proportions ?

(Haryana Board Hr. Sec., 1979)
[**Ans.** Yes]

LAW OF RECIPROCAL PROPORTIONS

1. Phosphine (PH_3) contains 91·11 per cent of phosphorus, water contains 88·79 per cent of oxygen, phosphorus pentoxide (P_2O_5) contains 43·66 per cent phosphorus. Show that these data illustrate the law of equivalent proportions for phosphorus and oxygen.

2. Ethylene (C_2H_4) contains 85·7% carbon, CO_2 contains 27·22 per cent carbon, water contains 11·11 per cent hydrogen. Show that these data illustrate the law of reciprocal proportions.

3. It is determined experimentally that one gram of hydrogen combines with 8 g of oxygen to form water. Also 2g of hydrogen combine with 32 g of sulphur to produce sulphuretted hydrogen. Sulphur and oxygen are found to combine with each other in proportions by weight of 1 : 1 or 2 : 3. Show that these illustrate one of the laws of chemical combination.

4. Carbon dioxide contains 27·27% carbon, carbon disulphide (CS_2) contains 15·79% carbon, sulphur tri-oxide contains 60% oxygen. Show that these results illustrate the law of equivalent proportions.

5. CO_2 contains 27·27% carbon, calcium carbide contains 37·5% carbon and CaO contains 28·57% oxygen. Show that the data illustrates the law of reciprocal proportions.

6. CS_2 contains 84·21% sulphur, S_2Cl_2 contains 47·4% sulphur and CCl_4 contains 7·79% carbon. Show that these results illustrate the Law of Reciprocal Proportions.

7. State the law of multiple proportions. Two chlorides of a metal contain respectively 64·2% and 47·2% of chlorine. How do these results illustrate the law of multiple proportions?
(Madurai Pre-Univ., 1975)

8. Two oxides of nitrogen M and N gave the following results; 4·4 g of M gave 2·24 litres of nitrogen and 6·0 g of N also gave 2·24 litres of nitrogen at N.T.P. Show by calculations, these results prove the law of multiple proportion.
(Gujarat Pre-Univ., 1975)

9. In four experiments the following percentage composition of hydrocarbons were obtained.

	Carbon	Hydrogen
(i)	75%	25%
(ii)	80%	20%
(iii)	85·7%	14·3%
(iv)	92·3%	7·7%

Show that these results illustrate the law of multiple proportions.
(Kashmir Pre-Univ., 1975)

10. Phosphine contains 91·17% phosphorus, water contains 83·80% of oxygen and an oxide of phosphorus (P_2O_5) contains 56·4% phosphorus. Show without using atomic weights that the data illustrates the law of reciprocal proportions. *(Guru Nanak Dev Pre-Univ., 1976)*

11. A metal M forms two oxides. One of the oxides contains 77·78% and the other 70% of the metal. Show that these values are in general agreement with the law of multiple proportions.
(All India Hr. Sec., 1977)

12. Phosphine contains 91·1% phosphorus and 8·9% hydrogen. Water contains 88·8% oxygen and 11·2% hydrogen. An oxide of phosphorus contains 56·4% phosphorus and 43·6% oxygen. Show that the above results illustrate the law of reciprocal proportions.
(Punjab Board Hr. Sec., 1977)

13. Show that following figures illustrate the law of chemical combination:

0·14g of an element ·A' is known to combine with 0·35g of element ·B' and 0·2857 g of element ·C' combines with 2·2857 g of element ·B'.
(Kurukshetra Pre-Univ., 1978)
State the Law.
[**Ans.** Law of Reciprocal Proportions]

14. Show that the following results illustrate the Law of Reciprocal Proportions:

(a) 0·46 g of magnesium on oxidation gave 0·77 g of magnesium oxide.

(b) 0·82 g of magnesium liberates 760 ml of hydrogen at N.T.P. from an acid (wt. of 1 ml of H_2 at N.T.P. $= 0·00009$ g).

(c) 1·11 g of oxygen react with hydrogen to give 1·25 g of water.
(*Punjab Board Hr. Sec., 1979*)

15. Hydrogen sulphide contains 94·11% sulphur ; sulphur dioxide contains 50% oxygen; and water contains 11·11% hydrogen. Show that the results are in agreement with the 'Law of Reciprocal Proportions'.
(*Kurukshetra Pre-Univ., 1974 ; Haryana Board Hr. Sec., 1978 ;
Panjab Pre-Univ., 1983*)

16. Show that the Law of Reciprocal Proportions is proved by the following figures :

1·4 g of an element A is known to combine with 1·6 g of element B while 0·5 g of another element C combine with 3·5 g of element A and 2·857 g of element C combine with 22·857 g of the element B.
(*Guru Nanak Dev Pre-Univ., 1981*)

3

Percentage Composition

The amount of an element in a compound can be found when the atomic weight of the element and the formula of the compound is known. The amount of an element in a compound can also be found by analysing a definite weight of the substance quantitatively. This amount is often expressed on the **per cent basis**, *which means the relative weights of the component elements present in one hundred parts by weight of the compound*. The term *per cent* is derived from the Latin expression, *percentum*, which literally means by the hundred or per 100. Thus if 1 part of an element is present in 100 parts by weight of the compound, we say 1 per cent or 1%. If it contains 22 parts by weight per 100 parts of the compound, it is expressed as 22 per cent or 22%. The percentange of an element in a compound is a fixed quantity.

DETERMINATION OF PERCENTAGE COMPOSITION

To determine the percentage composition of a compound when its chemical formula is given, we proceed as follows.

(1) *Multiply the atomic weight of an element by the number of atoms of that element present in the formula of the compound. This is to be done for each element in the compound.*

(2) *Find the sum total of all the weights found in step 1. This sum total is called the 'Molecular weight' of the compound or formula weight* of the compound. The **fraction of an**

Percentage Composition

element is the amount of the element present in 1 part by weight of the compound.

(3) Find the *fraction* of each element in the compound by dividing the total weight of that element by the molecular weight as found in step 2.

(4) Multiply the fraction of each element as found in step 3 by 100. The number so obtained is the per cent of each element present in the given compound.

Example 1. *Given the formula of magnesium pyrophosphate $Mg_2P_2O_7$; find the fraction and the per cent of each element in this compound.*

STEP I

Multiply the atomic weight of the elements Mg, P and O with the total no. of atoms present in the formula.

2 atoms of Mg = 2×24 = 48 parts by wt. of Mg
2 atoms of P = 2×31 = 62 ,, ,, ,, ,, ,, P
7 atoms of O = 7×16 = 112 ,, ,, ,, ,, ,, O

STEP II

Find the **Molecular weight** *of the compound.*

Mol. wt. = $48 + 62 + 112 = 222$.

STEP III

Find the fraction of each element by dividing the total weight of each element by Mol. wt.

Fraction of Mg = $\dfrac{48}{222}$ = 0.216

Fraction of P = $\dfrac{62}{222}$ = 0.279

Fraction of O = $\dfrac{112}{222}$ = 0.504

STEP IV

Find the **per cent** *of each element by multiplying the fraction of each element by 100.*

Per cent of Mg = $0.216 \times 100 = 21.6$
Per cent of P = $0.279 \times 100 = 27.9$
Per cent of O = $0.504 \times 100 = 50.4$

Example 2. *Calculate the percentage of CaO in $CaCO_3$*

Step I

Find the Molecular weight (or Formula weight) of $CaCO_3$.

1 atom of Ca = $1 \times 40 = 40$
1 atom of C = $1 \times 12 = 12$
3 atoms of O = $3 \times 16 = 48$
Formula wt. of $CaCO_3$ = 100

Step II

Find the formula weight of CaO

1 atom of Ca = $1 \times 40 = 40$
1 atom of O = $1 \times 16 = 16$
Formula wt. of CaO = 56

Step III

Find the fraction of CaO in $CaCO_3$.

$$\text{Fraction of CaO in CaCO}_3 = \frac{\text{Formula wt. of CaO}}{\text{Formula wt. of CaCO}_3}$$

$$= \frac{56}{100} = 0.56$$

Step IV

Find the per cent of CaO

Per cent of **CaO** = $0.56 \times 100 = 56$.

Example 3. *Calculate the percentage of water of crystallisation in Glauber's salt ($Na_2SO_4.10H_2O$).*

Step I

Find the formula weight of $Na_2SO_4.10H_2O$

2 sodium atoms = $2 \times 23 = 46$

$$\begin{aligned}
\text{1 sulphur atom} &= 1 \times 32 = 32 \\
\text{4 oxygen atoms} &= 4 \times 16 = 64 \\
\text{10 molecules of water} &= 10 \times 18 = 180 \\
\text{(Mol. wt. of } H_2O = 2+16 = 18) &= \text{———} \\
\text{Formula wt. of } Na_2SO_4.10H_2O &= 322
\end{aligned}$$

Step II

Find the fraction of water in $Na_2SO_4.10H_2O$.

$$\text{Fraction of water} = \frac{\text{Molecular wt. of water}}{\text{Formula wt. of } Na_2SO_4.10H_2O}$$

$$= \frac{180}{322} = 0.559$$

Step III

Find the per cent of water (of crystallisation)

$$\begin{aligned}
\text{Per cent of water} &= \text{fraction of water} \times 100 \\
&= 0.559 \times 100 \\
&= 55.9
\end{aligned}$$

Example 4. *A sample of impure FeS contains 50% Fe. Find the percentage of pure FeS in the sample.*

Step I

Find the formula weight of FeS.

$$\begin{aligned}
\text{1 atom of Fe} &= 1 \times 56 = 56 \\
\text{1 atom of S} &= 1 \times 32 = 32
\end{aligned}$$

Formula weight of FeS = 88

Step II

Find the fraction of Fe in FeS.

$$\text{Fraction of Fe} = \frac{\text{At. wt. of Fe}}{\text{Formula wt. of FeS}}$$

$$= \frac{56}{88} = 0.636$$

Step III

Find the per cent of Fe in FeS.

Per cent Fe in FeS = $0.636 \times 100 = 63.6$

Step IV

Find the percentage of pure FeS in the sample.

If the sample were 100% pure, the percentage of Fe would be 63·6

That is, if the percentage of Fe is 63·6, the percentage of pure FeS = 100

if the percentage of Fe is 1, the percentage of pure FeS

$$= \frac{100}{63 \cdot 6}$$

if the percentage of Fe is 50, the percentage of pure FeS

$$= \frac{100}{63 \cdot 6} \times 50$$

$$= 78 \cdot 61$$

Thus the percentage of pure FeS in the given sample is 78·61.

PERCENTAGE COMPOSITION FROM QUANTITATIVE ANALYSIS

Example 5. *A sample of clay was analysed. It contained 20% water. It was partially dried and the percentage of water in the partially dry sample was 8%. The percentage of alumina (Al_2O_3) in the partially dry sample was found to be 50%. What is the percentage of alumina in the original sample ?*

The partially dry sample contains water = 8 g

∴ Wt. of clay = 92 g

Thus, 92 g of clay contains alumina (Al_2O_3) = 50 g

In the original sample of clay, water = 20 g

∴ Wt. of residual clay = 100 − 20 = 80 g

Now, when the dry sample was 92 g, the wt. of Al_2O_3

$$= 50 \text{ g}$$

,, ,, ,, ,, ,, is 1 g, ,, $= \frac{50}{92}$ g

When the dry sample is 80 g, the weight of $Al_2O_3 = \frac{50}{92} \times 80$ g

$$= 43 \cdot 47 \text{ g}$$

Percentage Composition

Thus, the percentage of Al_2O_3 in the original sample
$$= 43.47 \text{ g}$$

Example 6. *A 2.91 g of a sample of silver metal was treated with nitric acid. On treatment with sodium chloride, the above solution gave 3.60 g of a precipitate of AgCl. Find the percentage purity of Ag in the original sample.*

Step I

Find the fraction of Ag in AgCl.

$$\text{Fraction of Ag in AgCl} = \frac{\text{At. wt. of Ag}}{\text{Formula wt. of AgCl}}$$
$$= \frac{108}{143.5} = 0.752$$

Step II

Find the number of grams of Ag in 3.60 g of AgCl.

Fraction of Ag in AgCl is 0.752 which means that **1 g** of AgCl contains Ag = 0.752 g.

3.60 g of AgCl will contain Ag = 3.60×0.752
$$= 2.707 \text{ g}$$

Thus, the sample of Ag 2.91 g contains pure Ag
$$= 2.707 \text{ g}$$

Step III

Find the percentage of Ag in the given sample.

Now, 2.91 g of the sample contains pure Ag = 2.707 g

$$1 \quad ,, \quad ,, \quad ,, \quad ,, \quad ,, \quad ,, \quad = \frac{2.707}{2.91} \text{ g}$$

$$100 \quad ,, \quad ,, \quad ,, \quad ,, \quad ,, \quad ,, \quad = \frac{2.707}{2.91} \times 100$$

$$= 93.02 \text{ g}$$

Thus the percentage purity of the sample = **93.02%**

Example 7. *A sample of coal contains 4% water. After drying, the moisture free residue contains 60% of carbon. Determine the percentage of carbon on the 'wet basis'.*

Wt. of water in the given sample = 4 g
∴ Wt. of dry sample = 100 − 4 = 96 g
Now wt. of carbon in 100 g of the dry sample = 60 g

,, ,, ,, ,, ,, 1 ,, ,, ,, ,, ,, ,, = $\dfrac{60}{100}$ g

and ,, ,, ,, ,, ,, 96 ,, ,, ,, ,, ,, ,,

(Or 100 g of the wet sample containing 4 g of moisture)

$$= \dfrac{60}{100} \times 96 \text{ g}$$
$$= 57 \cdot 2$$

Hence the percentage of carbon on the wet basis
= **57·2**

Example 8. *A sample of coal taken for analysis weighed 4·866 g. After heating and igniting the ash left behind weighed 0·149 g. Find the percentage of ash of this coal!*

The fraction of ash of the coal = $\dfrac{0\cdot 149}{4\cdot 866}$ = 0·0306

Percentage of ash of the coal = 0·0306 × 100
= **3·06%**.

Example 9. *0·69 g of sodium combines with 2·39 g of bromine to form sodium bromide. What is the percentage composition of sodium bromide?*

STEP I

Find the trial weight of sodium bromide.

Wt. of sodium = 0·69 g
Wt. of bromine = 2·39 g
∴ wt. of sodium bromide = 2·39 + 0·69
= 3·08 g

STEP II

Find the fraction of each element.

Fraction of sodium = $\dfrac{\text{wt. of sodium}}{\text{wt. of sodium bromide}}$

$= \dfrac{0\cdot 69}{3\cdot 08}$ = ·0224

Fraction of bromine $= \dfrac{\text{wt. of bromine}}{\text{wt. of sodium bromide}}$

$= \dfrac{2\cdot 39}{3\cdot 08} = 0\cdot 775.$

STEP III

Find the per cent of each element.

Per cent of sodium $= 0\cdot 224 \times 100 = 22\cdot 4\%$

Per cent of bromine $= 0\cdot 775 \times 100 = 77\cdot 5\%$

Example 10. *Methane contains 75% carbon and 25% hydrogen water contains 11·11% H. If C and O combine, what shall be the probable percentage composition of the compound ?*

(*Guru Nanak Dev Pre-Univ., 1980*)

In Methane (CH_4)

Percentage of C $= 75$

Percentage of H $= 25$

∴ Wt. of C which combines with 1 g of hydrogen

$= \dfrac{75}{25} = 3 \text{ g}$

In Water (H_2O)

Percentage of H $= 11\cdot 11$

Percentage of O $= 100 - 11\cdot 11 = 88\cdot 89.$

∴ Wt. of oxygen which combines with 1 *g* of H

$= \dfrac{88\cdot 89}{11\cdot 11} = 8 \, g$

∴ the ratio of wts. of C and O which **separately combine** with a fixed wt. of H (1 *g*) $= 3 : 8$

Also if C and O combine with each other-the ratio of wts. should be 3 : 8.

∴ probable percentage composition of the compound of carbon and oxygen is :

Percentage of carbon $= \dfrac{3}{11} \times 100 = 27\cdot 27$

and ,, ,, oxygen $= \dfrac{8}{11} \times 100 = 72\cdot 73.$

END-OF-CHAPTER PROBLEMS

1. Find the formula weight of the following :

   (a) Water (H_2O).
   (b) Ferrous sulphide (FeS).
   (c) Magnesium sulphate.
   (d) Sodium hydroxide.
   (e) Sodium chloride.
   (f) Magnesium chloride.
   (g) Mercuric oxide.
   (h) $CuSO_4.5H_2O$.
   (i) $K_2SO_4.Al_2(SO_4)_3.24H_2O$.
   (j) $FeSO_4.7H_2O$.

   [**Ans.** (a) 18; (b) 88; (c) 120; (d) 40; (e) 58·5; (f) 95; (g) 216·61; (h) 249·71; (i) 948·702; (j) 278].

2. Find the fraction of each of the elements present in

   (a) $CaCO_3$ (b) CH_4 (c) CO (d) CO_2

   [**Ans.** (a) Ca = 0·4;   (b) C = 0·75;
   C = 0·12;   H = 0·25;
   O = 0·48;]

   (c) C = 0·428;   (d) C = 0·2727
   O = 0·572;   O = 0·7348]

3. Find the fraction of (a) CaO in $Ca_3(PO_4)_2$ (b) MgO in $Mg(OH)_2$ (c) P_2O_5 in $Ca_3(PO_4)_2$.

   [**Ans.** (a) CaO = 0·542; (b) MgO = 0·691; (c) P_2O_5 = 0·47]

4. Find the fraction of (a) KCl in $KCl.MgCl_2.6H_2O$; (b) H_2O in $MgSO_4.7H_2O$.

   [**Ans.** (a) KCl = 0·268; (b) H_2O = 0·512]

5. Determine the percentage of iron in each of the following compounds : (a) $FeCO_3$ (b) Fe_2O_2 (c) Fe_3O_4

   [**Ans.** (a) 48·2%; (b) 69·94%; (c) 72·37%]

6. Determine the percentage of copper and oxygen in each of the following oxides : (a) CuO (b) Cu_2O).

   [**Ans.** (a) Cu = 79·74; O = 20·26
   (b) Cu = 83·4; O = 11·6]

7. Find the percentage composition of (a) K_2CO_3 (b) K_2SO_4.

   [**Ans.** (a) %ge of K = 56·58; C = 8·59; O = 34·73
   (b) %ge of K = 44·88; S = 18·37; O = 36·73]

8. Determine the percentage of chromium trioxide (CrO_3) in (a) Potassium chromate (K_2CrO_4) and (b) Potassium dichromate ($K_2Cr_2O_7$).

   [**Ans.** (a) 51·5%; (b) 67·9%]

9. Determine the percentage composition of dolomite ($CaCO_3.MgCO_3$).

   [**Ans.** %ge of Ca = 21·73; %ge of Mg = 13·18;
   %ge of C = 13·01; %ge of O = 52·08]

Percentage Composition

10. Determine the percentage of water of crystallisation in
    (a) $FeSO_4.7H_2O$.
    (b) $MgSO_4.7H_2O$.
    (c) $K_2SO_4.Al_2(SO_4)_3.24H_2O$.

    [**Ans.** (a) 45·32% (b) 51·2% (c) 45·56%]

11. Determine fraction and percentage of nitrogen in (a) NH_4NO_3. (b) $(NH_4)_2SO_4$.

    [**Ans.** (a) 0·35, 35% ; (b) 0·212, 21·2%]

12. What weight of silver is present in 6·90 g of Ag_2S?

    [**Ans.** 6·21 g]

13. A clay contains 45% alumina (Al_2O_3) and 10% water. What is the percentage of alumina on the 'dry' basis ?

14. How much phosphorus is contained in 10 g of the compound $CaCO_3.3Ca_3(PO_4)_2$?

    [**Ans.** 1·99]

15. What amount of sulphur is required to prepare 500 g of H_2SO_4 ?

    [**Ans.** 163·2 g]

16. $CaSO_4.2H_2O$ is heated so as it loses only three-fourths of its water content. (a) What is the percentage loss of weight ? (b) What weight of gypsum must be heated to make 50 g of sample containing only 25% water ?

    [**Ans.** (a) 15·69%; (b) 118·3 g of $CaSO_4.2H_2O$]

17. One gram of potassium chlorate gave on analysis 0·3919 g of potassium, 0·2892 g of chlorine and 0·314 g of oxygen. Determine the percentage composition of potassium chlorate.

    [**Ans.** K = 31·89%; Cl = 28·92%; O = 39·19]

18. A sample of impure cuprous oxide (Cu_2O) contains 56% copper. What is the percentage of pure Cu_2O in the sample ?

    [**Ans.** 62·7%]

19. What weight of CuO will be required to prepare 300 grams of copper ?

    [**Ans.** 375·5]

20. Zinc metal is obtained from the ore ZnS. What weight of zinc metal will be obtained from 1200 g of the ore which is only 83% pure.

21. Calculate the percentage of copper in copper sulphate crystals ($CuSO_4.5H_2O$).

    [**Ans.** 25·8%]

22. Methane contains 75% carbon and 25% hydrogen. Water contains 11·11% H. If C and O combine, what shall be the probable percentage composition of the compound ?

    (*Guru Nanak Dev Pre-Univ., 1980*)

    [**Ans.** Percentage of C = 27·27 ;
    Percentage of O = 72·73]

4

Mole Concept

Since the atomic weight of an element is proportional to the actual weight of an element, it is customary to express the atomic weight of an element in grams. Thus the atomic weight of an element expressed in grams is called **Gram-Atomic Weight** or simply the **gram-atom** (g-atom). Similarly the molecular weight of any given substance is simply the molecular weight of the substance in grams. This is called the **Gram-Molecular Weight** or **gram-mole** of the substance. Since it is extensitively used in chemical calculations, the simple word **Mole** *has been coined as an abbreviation of the term gram mole.* Thus when we speak 1 gram atom of chlorine, we mean 35·5 g of chlorine and when we say 1 gram-mole or 1 mole of chlorine, we mean 71 g of chlorine.

DEFINITION OF MOLE

Mole has been defined differently at different occasions depending upon the nature of the substance under consideration *i.e.*, gaseous form of the substance, ionic form, molecular form, or atomic form. Whatever the case, the word mole must be followed by the terms, atom, molecule, litres, electrons, protons etc., in order to represent the nature of the substance under consideration. Thus mole may be defined as :

1. The weight of the substance in grams which contains Avogadro number of *atoms* of the substance (*i.e.*, $6·023 \times 10^{23}$ atoms). It is also called **gram atomic**

Mole Concept

weight or 1-gram atomic weight of the substance (usually an element).

2. The weight of the substance in grams which contains Avogadro number of *molecules* of the substance (6.023×10^{23} molecules). It is also called the **gram molecular weight** of the substance (element or compound).

3. The weight of 22.4 litres of any gas or vapours at NTP. This volume is also called **molar volume** of the substance.

4. The gram formula weight of any substance (usually ionic compounds).

5. The Avogadro number of charged particles in case of sub-atomic particles such as electrons, protons and neutrons.

Example 1. *What is the weight of 3 g-atoms of sulphur ?*

1 gram-atom of sulphur has 32 grams

3 ,, ,, ,, ,, ,, $3 \times 32 = 96$ grams

Example 2. *How many gram-atoms are present in 144 g of sulphur ?*

1 gram-atom of sulphur weighs $= 32$ g

or 32 g of sulphur make 1 g-atom of sulphur

1 ,, ,, ,, $\frac{1}{32}$ g-atom of sulphur

144 ,, ,, ,, $\frac{1}{32} \times 144 = 4.5$ g-atoms

Example 3. *How many grams of calcium are present in 4.25 g-atoms of calcium ?* (At. wt. of Ca = 40).

1 gram-atom calcium has 40 g

∴ 4.25 gram-atoms of calcium will have 40×4.25

$= 170.00$ g

Example 4. *What is the weight of 3.5 moles of $KClO_3$?*

Step I

Find the molecular weight of $KClO_3$.

The molecular weight of $KClO_3$ is found as follows:

1 atom wt. of K	$= 1 \times 39$	$= 39$
1 atom wt. of Cl	$= 1 \times 35 \cdot 5$	$= 35 \cdot 5$
3 atoms wt. of O	$= 3 \times 16$	$= 48$
Molecular weight of $KClO_3$		$= 122 \cdot 5$

Step II

Find the wt. of 3·5 moles of $KClO_3$.

Since the molecular weight of $KClO_3$ is 122·5, we can say

1 mole of $KClO_3$ weighs $= 122 \cdot 5$ g

3·5 moles of $KClO_3$ will weigh $= 3 \cdot 5 \times 122 \cdot 5$
$= 428 \cdot 75$ g

Example 5. *How many moles of* $CaCO_3$ *are contained in* 400 g *of* $CaCO_3$?

Step I

Find the molecular weight of $CaCO_3$.

The molecular weight of $CaCO_3$ is

$1 \times$ at. wt. of Ca	$= 1 \times 40$	$= 40$
$1 \times$ at. wt. of C	$= 1 \times 12$	$= 12$
$3 \times$ at. wt. of O	$= 3 \times 16$	$= 48$
	Mol. wt.	$= 100$

Step II

Find the number of moles in the given wt. of calcium carbonate.

Since the molecular weight of $CaCO_3$ is 100, we can say that

100 g of $CaCO_3$ are $= 1$ mole

1 ,, ,, ,, $= \dfrac{1}{100}$

440 ,, ,, ,, $= \dfrac{1}{100} \times 400 = 4$ moles

Example 6. *How many grams and gram-atoms of sulphur are contained in 0·25 mole of H_2SO_4.*

STEP I

Find the number of grams in 0·25 mole H_2SO_4.

The molecular weight of H_2SO_4 $= 2 + 32 + 64 = 98$
Thus, 1 mole of H_2SO_4 weighs $= 98$ g
and 0·25 mole of H_2SO_4 will weigh $= 98 \times 0.25$
$= 24.50$ g

STEP II

Find the number of grams of sulphur in 24·50 g of H_2SO_4 (or 0·25 mole of H_2SO_4).

1 mole (or 98 g) of H_2SO_4 contains S $= 32$ g
0·25 mole (or 24·5 g) will contain S $= 32 \times 0.25$
$= 8.00$ g

$$\left(\text{or } \dfrac{32}{98} \times 24.5 = 8.00 \text{ g} \right)$$

STEP III

Find the gram-atom of sulphur in 0·25 mole of H_2SO_4.

1 mole of H_2SO_4 has S $= 1$ g-atom
∴ 0·25 ,, ,, will have S $= 1 \times 0.25$
$= 0.25$ mole

Example 7. *In a certain reaction 0·5 mole of $NaHCO_3$ were required. Express this quantity in grams.*

STEP I

Find the molecular weight of $NaHCO_3$.

3—NPC

$$
\begin{aligned}
1\ Na &= 1 \times 23 = 23 \\
1\ H &= 1 \times 1 = 1 \\
1\ C &= 1 \times 12 = 12 \\
3\ O &= 3 \times 16 = 48 \\
\therefore \text{Molecular wt.} & = 84
\end{aligned}
$$

Step II

Now 1 mole of $NaHCO_3$ weighs = 84 g

0.4 ,, ,, will weigh = $84 \times 0.4 = 33.6$ g

NUMBER OF ATOMS OR MOLECULES IN MOLES

According to Avogadro's number we know that the **gram atomic weight or gram-atom of any element contains the same number of atoms**. It has been shown by Avogadro that the number of atoms present in 1 gram-atom of an element is 6.02×10^{23}. This means that:

1 gram-atom of sulphur will have 6.02×10^{23} atoms of sulphur or 1 gram-atom of sodium will have 6.02×10^{23} atoms of sodium.

Example 8. *How many atoms of hydrogen and sulphur are present in 0.8 mole of H_2S?*

From the Avogadro's number, we have:

1 mole of H_2S has hydrogen atoms = $6.02 \times 10^{23} \times 2$

$\therefore$ 0.8 ,, ,, will have ,, = $2 \times 0.8 \times 6.02 \times 10^{23}$
$\phantom{\therefore 0.8 \text{ ,, ,, will have ,, }} = 9.632 \times 10^{23}$ atoms

Similarly

1 mole of H_2S has sulphur atoms = 6.02×10^{23}

0.8 ,, ,, will have ,, ,, = $0.8 \times 10^{23} \times 6.02$
$\phantom{0.8 \text{ ,, ,, will have ,, ,, }} = 4.816 \times 10^{23}$ atoms

Example 9. *How many molecules of SO_2 are present in 0.4 mole of SO_2?*

1 mole of SO_2 will have = 6.02×10^{23} molecules of SO_2

$\therefore$ 0.4 mole of SO_2 ,, ,, = $0.4 \times 6.02 \times 10^{23}$
$\phantom{\therefore 0.4 \text{ mole of } SO_2 \text{ ,, ,, }} = 2.408 \times 10^{23}$ molecules of SO_2.

Mole Concept

Example 10. *How many atoms and molecules of phosphorus are present in 93 grams of phosphorus ?*

Step I

Find the number of gram-atoms of phosphorus.

The atomic weight of phosphorus is 31

$\therefore$ 31 g of phosphorus make gram-atoms = 1

$\therefore$ 93 ,, ,, ,, ,, ,, ,, $= \dfrac{1}{31} \times 93$

= 3 g-atoms

Step II

Find the number of atoms.

According to Avogadro's number the number of atoms in 1 g-atom of phosphorus $= 6.02 \times 10^{23}$

$\therefore$ The number of atoms in 3 g-atom

$= 3 \times 6.02 \times 10^{23}$

$= 18.06 \times 10^{23}$ atoms

or $= 1.806 \times 10^{24}$ atoms

Step III

Find the number of phosphorus.

The molecular weight of phosphorus $(P_4) = 31 \times 4 = 124$

Thus 124 g of phosphorus are = 1 mole

93 ,, ,, ,, $= \dfrac{1}{124} \times 93$

= 0.75 mole

Step IV

Find the number of molecules of phosphorus.

According to Avogadro's number,

1 mole of phosphorus has molecules $= 6.02 \times 10^{23}$

$\therefore$ 0.75 ,, ,, ,, will have ,, $= 0.75 \times 6.02 \times 10^{23}$

$= 4.74 \times 10^{23}$ moles.

Example 11. (a) *Which of the following weighs maximum?*

(i) *50 g of iron,* (ii) *5 g atoms of nitrogen,* (iii) *0·1 atom of silver,* (vi) *1×10^{23} atoms of carbon*

(b) *How many atoms and gram atoms are there in 10 g of calcium?*

(*Himachal Pre-Univ., 1981*)

(a) (i) Wt. of iron = $50\ g$

(ii) 1 g-atoms of nitrogen weighs = $14\ g$ (at. wt. of nitrogen)

∴ 5 g-atoms of nitrogen will weigh = $14 \times 5 = 70$ g

(iii) 1 g-atom of Ag weighs 108 g

∴ 0·1 g-atom of Ag will weigh = $108 \times 0·1 = 10·8$ g

(iv) Now 1 g-atom of C = $6·023 \times 10^{23}$ atoms of carbon

Also, at. wt. of carbon = 12

∴ $6·023 \times 10^{23}$ atoms of carbon = $12\ g$

∴ 1×10^{23} atoms of carbon = $\dfrac{12}{6·023 \times 10^{23}} \times 1 \times 10^{23}$

= $1·992$ g

Thus we conclude that 5 g-atoms of nitrogen weigh the maximum.

(b) The at. wt. of Ca = 40

∴ 1 g-atom of Ca = $40\ g$

Now 40 g of Ca = 1 g-atoms

∴ 10 g of Ca = $6·023 \times 10^{23}$ atoms of Ca

10 g of Ca = $\dfrac{6·02 \times 10^{23}}{40} \times 10$

= $1·506 \times 10^{23}$ **atoms.**

Example 12. *What is the volume occupied by 2·5 moles of carbon dioxide at N.T.P.?* (*Punjabi Pre-Univ., 1981*)

22·4 litres of CO_2 at N.T.P. weigh equal to 1 g-mole or 1 mole.

This means that the volume at N.T.P. occupied by one mole of CO_2 = 22·4 ltires.

∴ The volume occupied by 2·5 moles of CO_2

$$= \frac{22·4}{1} \times 2·5$$

$$= 56 \text{ litres}$$

Example 13. *Calculate the volume at N.T.P. occupied by*

(i) *0·5 moles N_2*

(ii) *$6·023 \times 10^{23}$ molecules of H_2S*

(iii) *one gram equivalent weight of oxygen.*

(*Punjab Pre-Univ., 1980*)

(i) The volume occupied by 1 mole of nitrogen at N.T.P.
= 22·4 litres

∴ the volume ,, 0·5 mole of ,, = $22·4 \times 0·5$
= 11·1 litres

(ii) $6·023 \times 10^{23}$ molecules of H_2S occupy volume (at N.T.P.)
= 22·4 litres

∴ $6·023 \times 10^{22}$ molecules of H_2S will occupy volume (at N.T.P.) = $\dfrac{22·4}{6·023 \times 10^{23}} \times 6·023 \times 10^{22}$

= 2·24 litres.

(iii) One g-equivalent weight of oxygen = 8 g

Now we know that :

32 g of oxygen (1 mole) occupy volume at N.T.P.
= 22·4 litres

∴ 8 g of oxygen will occupy volume at N.T.P.

$$= \frac{22·4}{32} \times 8 = 5·6 \text{ litres.}$$

END-OF-CHAPTER PROBLEMS

1. Find the number of gram-atoms of the elements contained in each of the following :

(a) 206 g of Ba (b) 415·45 of Sn (c) 11·45 g of zinc

(d) 0·45 g of C (e) 69·0 g of sodium (f) 65·5 g of Al

(g) 80·0 g of oxygen   (h) 80·0 g of sulphur

[**Ans.** $a = 1·5$ g-atom ; $b = 3·5$ g-atom ;
$c = 1·75$ g-atom ; $d = 0·0375$ g-atom ;
$e = 3·0$ g-atom ; $f = 2·5$ g-atom ;
$g = 5·0$ g-atom ; $h = 2·5$ g-atom]

2. How many grams are contained in :

(a) 0·4 atom of carbon   (b) 0·5 g-atom of sodium
(c) 0·8 g-atom of sulphur   (d) 0·6 g-atom of magnesium
(e) 0·2 g-atom of phosphorous   (f) 0·9 g-atom of zinc.

[**Ans.** $a = 4·8$ g ; $b = 11·5$ g ; $c = 25·6$ g ;
$d = 14·4$ g ; $e = 6·2$ g ; $f = 58·5$ g]

3 Calculate the number of grams in *one* mole of each of the following substances :

(a) C_2H_4 (b) Na_2CO_3 (c) $NaOH$ (d) KOH (e) $CaCO_3$
(f) $FeSO_4 \cdot 7H_2O$ (g) $MgCl_2$ (h) $KMnO_4$ (i) HCl.

[**Ans** $a = 28$ g ; $b = 106$ g ;
$c = 40$ g ; $d = 56$ g ;
$e = 100$ g ; $f = 278$ g ;
$g = 95$ g ; $h = 158$ g ;
$i = 36·5$ g]

4. How many gram-moles are contained in 50 g of calcium carbonate ?   [**Ans.** 0·50 g-mole]

5. How many gram-moles (moles) are contained in each of the following :

(a) 50 g of $CaCO_3$ (b) 20 g of $NaOH$
(c) 31·6 g of $KMnO_4$ (d) 15·2 g of $FeSO_4$
(e) 49 g of H_2SO_4 (f) 80 g of oxygen
(g) 124 g of phosphorus   (h) 71 g of chlorine
(i) 28 g of nitrogen   (j) 2 g of hydrogen

[**Ans.** $a = 0·50$ mole ; $b = 0·5$ mole ;
$c = 0·02$ mole $b = 0·5$ mole
$e = 0·5$ mole ; $f = 2·5$ mole
$g = 1·00$ mole ; $h = 1·00$ mole
$i = 1·00$ mole ; $j = 1·00$ mole]

6. A phosphate rock was found to contain 42% $Ca_3(PO_4)_2$ by weight. How many moles of $Ca_3(PO_4)_2$ are contained in 1 kilogram (1000 g) of the rock ?   [**Ans.** 19·7 moles]

7. A 500 ml solution of hydrochloric acid in water contains 18·25 g of HCl. How many moles of HCl are present in this solution ? How many moles will be needed to prepare 1500 ml solution of the same strength ?   [**Ans.** 0·5 mole; 1·5 moles]

Mole Concept

8. How many atoms of each kind are present in :

(a) 0.4 mole of SO_2

(b) 0.5 mole of $CaCO_3$

(c) 0.8 mole of NaCl

(d) 0.5 mole of SO_3.

[**Ans.** $a = 2.408 \times 10^{23}$ atoms of each of S and double of O;

$b = 3.01 \times 10^{23}$ atoms each of Ca, C and three times of O]

$c = 4.816 \times 10^{23}$ atoms of each Na and Cl;

$d = 3.010 \times 10^{23}$ atoms of each S and three times of O]

9. How many molecules of the following are present in :

(a) 1.0 mole of $KMnO_4$

(b) 0.6 mole of HCl

(c) 0.7 mole of NaOH

(d) 0.5 mole of KCl

(e) 5.0 mole of Cl_2.

[**Ans.** $a = 6.02 \times 10^{23}$ molecules of $KMnO_4$;

$b = 3.613 \times 10^{23}$ molecules of HCl;

$c = 4.216 \times 10^{23}$ molecules of NaOH;

$d = 3.010 \times 10^{23}$ molecules of KCl;

$e = 30.10 \times 10^{23}$ molecules of Cl_2]

10. (a) How many grams of calcium carbonate be heated to get enough carbon dioxide which can convert 0.1 mole of sodium carbonate into sodium bicarbonate ?

[**Ans.** 0.1 mole of $CaCO_3$]

11. (a) Which of the following weighs maximum :

(i) 50 g of iron

(ii) 5 g atom of nitrogen

(iii) 0.10 gram atoms of silver

(iv) 1×10^{23} atoms of carbon ?

[**Ans.** (ii) 5g atom of nitrogen]

(b) How many atoms and how many gram atoms are there in 10 g of calcium ?

(At. wt. of N = 14, C = 12, Ca = 40]

[**Ans.** 1.506×10^{23} atom]

1.506×10^{23} g atom]

12. (i) How many grams of ammonia are contained in its 0.56 mole ?

(ii) How many gram atoms of sulphur and hydrogen are contained in 1.5 mole of hydrogen sulphide ?

(iii) How many grams of carbon and oxygen are present in 2.5 mole of carbon dioxide ?

(iv) How many chlorine molecules are there in 71 gm of chlorine gas at S.T.P. ?

(Panjab Pre-Univ., 1974)

[**Ans.** (i) 9·52   (ii) 3, 1·5   (iii) 30, 80
(iv) $6·023 \times 10^{23}$]

13. Calculate the number of molecules in 150 ml of oxygen at 20°C and 750 mm pressure.

(Kurukshetra Pre Univ., 1974)

[**Ans.** $3·71 \times 10^{21}$]

14. (a) How many atoms are there in 0·5 moles of sodium ?
(b) Which of the following weighs most and why ?
  (i) 5·0 g of nitrogen
    (ii) 5·0 g-atoms of nitrogen
    (iii) 5·0 g-molecules of nitrogen
    (iv) 5·0 molecules of nitrogen.

(Panjab Pre Univ., 1976)

[**Ans.** (iii) 5g molecules of nitrogen]

15. Which mole is heavier and why — one mole of Na or one mole of oxygen gas ?   *(Kuruksketra Pre-Univ., 1976)*

[**Ans.** one mole of oxygen]

16. Which of the following weigh maximum and which minimum :

  (i) 18 g of nitrogen;   (ii) 1·5 g-atom of oxygen ;
  (iii) 1·6 moles of magnesium :   (iv) $3·01 \times 10^{23}$ atoms of iron

[At. wt. of N = 14, O = 16, Mg = 24, Fe = 56]

(Himachal Pre Univ., 1976)

[**Ans.** Maximum = 1·6 moles of Mg]
Minimum = 18 g of nitrogen]

17. What weight of sodium (At. wt. 23) contains the same number of atoms as 3 g of carbon ?   *(All India Hr. Sec. 1977)*

[**Ans.** 5·75 g]

18. What weight of nitrogen contains the same number of molecules as 11·2 litres of carbon dioxide ?   *(All India Hr. Sec. 1977)*

[**Ans.** 14 g]

19. Calculate the following :
(a) Number of moles in 18 g of glucose ($C_5H_{12}O_6$)
(b) Number of moles in 18 g of water (H_2O),
(c) Number of moles in 0·32 g of oxygen.
(d) Weight of one molecule of nitrogen.
(e) Weight of 2·24 litres of CO_2 at N.T.P.

(Panjab Board Hr. Sec , 1977)

[**Ans.** (a) = 0·1 mole;   (b) = 1 mole
(c) = 0·01 mole;   (d) = $4·63 \times 10^{-23}$ g
(e) = 4·4 g]

20. (a) Calculate the actual weight of a single atom of hydrogen.

(b) What is the weight of 2·5 moles of H_2SO_4.

(Himachal Pre-Univ., 1978)

[**Ans.** (a) $0·166 \times 10^{-23}$ g

(b) 245 g]

21. Calculate the volume at N.T.P. occupied by 4 g of oxygen.
(Himachal Pre-Univ., 1978)

[**Ans.** 2800 cc]

22. Atomic weight of an element is 22·99 a.m.u. How many atoms and gram atoms are in 92 grams of the element ?
(Punjab Pre-Univ., 1978)

[**Ans.** No. of atoms = $24·09 \times 10^{23}$]

No. of gram atoms = 4·001]

23. (a) What is the significance of the figure $6·023 \times 10^{23}$ in chemical calculations ?

(b) Calculate the number of atoms in

(i) 1·0 g of Hydrogen

(ii) 1·0 g of Carbon

(iii) 1·0 g of Aluminium

(iv) 1·0 g of Sulphur.

(c) Taking the example of Iron, show which has the greatest mass : one atom; one gram; or one gram atom.

[H = 1; C = 12; Al = 27; S = 32; Fe = 56]
(Kurukshetra Pre-Univ., 1978)

[**Ans.** (b) (i) $6·02 \times 10^{23}$;   (ii) $0·5 \times 10^{23}$

(iii) $0·22 \times 10^{23}$; and   (iv) $0·19 \times 10^{23}$

(c) One gram atom of Fe = 56 g]

24. How many atoms of various elements are indicated by the following formulae :

(a) 6 $CuSO_4$ $5H_2O$

(b) 12 $Mg_3(PO_4)_2$.

(Guru Nanak Dev Pre-Univ., 1978)

Ans. (a) No. of Cu atoms = 6; No. of S atoms = 6;

No. of H atoms = 60; and No. of O atoms = 54

(b) No. of Mg atoms = 36; No. of P atoms = 24 ;
and No. of O atoms = 96]

25. Calculate the actual volume of one molecule of water (liquid).

[Mol. wt. of H_2O = 18; Density of H_2O = 1; and

Avogadro's Number (N) = $6·02 \times 10^{23}$]

(Punjabi Pre-Univ., 1971)

[**Ans.** $2·99 \times 10^{-23}$ cc]

26. Fill in the blanks :

$$Zn(s) + H_2SO_4\ (l) \longrightarrow ZnSO_4(z) + H_2(g)$$

(a) Moles : 2 moles + ⟶ +

(b) Grams ; 130·74 + ⟶ +

(Haryana Board Hr. Sec., 1979)

[**Ans.** (a) 2 moles + 2 moles → 2 moles + 2 moles
(b) 130·74 g + 197·11 g → 323·83 g + 4·02 g]

27. Complete the following sentences :

(a) 0·5 mole of CO_2 contains—molecules.

(b) 10 00 g of Ca contains—atoms.

(c) N.T.P. stands for—°C—pressure.

(Haryana Board Hr. Sec., 1976)

[**Ans.** (a) $3·01 \times 10^{23}$
(b) $1·50 \times 10^{23}$
(c) 0; 760 mm]

28. Calculate the number of moles in each of the following amounts of materials :

(i) 10 g of $CaCO_3$

(ii) 1×10^{23} molecules of CO_2

(At. Wts. of Ca = 40; C = 12; O = 16)

(Himachal Pre-Univ., 1980)

[**Ans.** (i) 0·1 mole (ii) 0·166 moles]

29. Calculate the weight of $18·06 \times 10^{23}$ molecules of carbon monoxide. *(Panjabi Pre Univ., 1980)*

[**Ans.** 84 g]

30. What is the number of molecules in 22·6 litres of oxygen at STP. *(Punjab Pre-Univ., 1981)*

[**Ans.** $6·023 \times 10^{23}$ molecules]

5

Empirical and Molecular Formulae

We have seen that the percentage composition of a compound can be calculated from the formula when it is known. It can also be found from the composition as determined by actual quantitative analysis. To derive the chemical formula of a compound from its percentage composition, it is necessary to determine:

(i) *The relative number of atoms in the molecule of the substance by dividing the weight per cent of each element by its atomic weight.* The quotient so obtained will be the number of gram-atoms of the substance in 100 g of the substance.

(ii) *Express the number of gram-atoms of each element in the ratio of the lowest whole-number integers.* These whole numbers are the relative number of atoms in the molecule.

The formula so obtained is termed as the Empirical Formula of a molecule of the substance. Thus Empirical formula of a substance is the *simplest* formula of a compound. It is not necessarily the molecular formula of a compound. Thus **Empirical formula** *of a compound is the simplest formula which expresses the simplest whole number ratio between the atoms of various elements present in a compound.*

Example 1. *A compound of iron has 63·52% Fe and 35·48% sulphur. Determine the empirical formula of the compound.*

43

Step I

Find the gram-atoms of Fe and S by dividing their per cent weight by their respective atomic weights.

gram-atoms of Fe $= \dfrac{63 \cdot 52}{56} = 1 \cdot 134$

gram-atoms of S $= \dfrac{35 \cdot 48}{32} = 1 \cdot 125$

Step II

Reduce the above ratio to simple whole numbers by dividing each gram by the lowest value.

$$Fe = \dfrac{1 \cdot 134}{1 \cdot 134} = 1$$

$$S = \dfrac{1 \cdot 135}{1 \cdot 134} = 1 \text{ (approx.)}$$

Step III

Place these whole number below the symbol of the respective element on the lower right hand corner.

Thus the empirical formula of the substance is

$Fe_1 S_1$ or FeS.

Example 2. *The analysis of a compound of manganese and oxygen show that 7·2026 g of manganese combines with 2·7974 g of oxygen. Determine the empirical formula of the compound.*

Step I

Determine the gram-atom of each element

(Ratio of the no. of atoms)

gram-atoms of Mn $= \dfrac{\text{wt. of Mn}}{\text{At. wt. of Mn}}$

$= \dfrac{7 \cdot 2026}{25} = 0 \cdot 1309$

gram-atoms of oxygen $= \dfrac{\text{wt. of oxygen}}{\text{At. wt. of oxygen}}$

$= \dfrac{2 \cdot 7974}{16} = 0 \cdot 1748$

Empirical and Molecular Formulae

STEP II

Divide the above ratio of gram atoms by the lowest ratio.

$$Mn = \frac{0.1309}{0.1309} = 1$$

$$O = \frac{0.1748}{0.1309} = 1.33$$

STEP III

Multiply the above ratio by 3 to get whole numbers.

$Mn = 1 \times 3$

$O = 1.33 \times 3 = 3.99 = 4$ approx.

STEP IV

Place the whole numbers below the symbols of the respective elements on the lower right hand corner.

Thus empirical formula of the compound

$$= Mn_3O_4$$

Example 3. *The percentage composition of Potassium ferrocyanide is $K = 42.4\%$, $Fe = 15.2\%$, $CN = 42.4\%$. Determine the empirical formula.*

The empirical formula may be determined by the same steps shown below in a tabular form.

Element or radical	Percentage of element or radical	At. wt. of element or formula wt. of the radical	Ratio of the No. atoms or radical	Smallest whole number ratio of atoms or radicals	Nearest whole number
K	42.4	39.0	$\frac{42.4}{39.0} = 1.08$	$\frac{1.08}{0.272} = 3.97$	4
Fe	15.2	56	$\frac{15.2}{56} = 0.272$	$\frac{0.272}{0.272} = 1.00$	1
CN	42.4	26	$\frac{42.4}{26} = 1.63$	$\frac{1.63}{0.272} = 5.99$	6

$\therefore$ Empirical formula $= K_4[Fe(CN)_6]$

Example 4. 1.425 g of a sample of organic compound was burnt completely giving CO_2 and water only. The weights of CO_2 and water obtained were 1.771 g and 0.725 g respectively. The compound contains C, H and O only. What is the empirical formula of the compound?

STEP I

Find the percentage of C and H from given data.

Molecular weight of $CO_2 = 12+32 = 44$

and Mol. wt. of water $= 2+16 = 18$

We know that 44 g of CO_2 contain C $= 12$ g

$\therefore$ 1 g of CO_2 contains C $= \dfrac{12}{44}$ g

and 1.771 g of CO_2 contain C $= \dfrac{12}{44} \times 1.771$

$= 0.483$ g

Now, the wt. of carbon in 1.771 g CO_2 or 1.425 g of the compound is 0.483.

$\therefore$ the fraction of carbon in the compound $= \dfrac{0.483}{1.425}$

$= 0.3388$

and the percentage of carbon $= 0.3388 \times 100 = 33.88$

Also, we know that 18 g of water contain

hydrogen $= 2$ g

$\therefore$,, ,, 1 ,, , $= \dfrac{2}{18}$

and ,, ,, 0.725 ,, ,, $= \dfrac{2}{18} \times 0.725$

$= 0.0805$ g

Now, the wt of hydrogen in 0.725 g of water or 1.425 g of the compound is 0.0805 g.

$\therefore$ the fraction of hydrogen in the compound $= \dfrac{0.0805}{1.425}$

$= 0.0565$

$\therefore$ the percentage of hydrogen $= 0.0565 \times 100$

$= 5.65$

Empirical and Molecular Formulae

STEP II

Find the percentage of oxygen by substracting the sum total of the percentage of C and H from 100.

Thus, percentage of $O = 100-(33\cdot88+5\cdot65)$
$= 60\cdot57$

STEP III

Determine the empirical formula by usual steps.

Element	Percentage of each element	At. wt. of each element	Ratio of the no. of atoms	Smallest whole no. ratio	Simplest whole no. ratio
C	33·88	12	2·823	$\frac{2\cdot823}{2\cdot823}=1$	$1\times3=3$
H	5·65	1	5·65	$\frac{5\cdot65}{2\cdot823}=2$	$2\times3=6$
O	60·47	16	3·78	$\frac{3\cdot78}{2\cdot823}=1\cdot34$	$1\cdot34\times3=4$

Thus the empirical formula of the compound is $C_3H_6O_4$.

Example 5. *A double sulphate of potassium and aluminium contains $K = 8\cdot24\%$, $Al = 5\cdot71\%$, $SO_4 = 40\cdot49\%$. What is the empirical formula of the compound?*

Proceeding in the same way.

Element or radical	Percentage of the element or radical	At. wt. of the element or mol. wt. of the radical	Ratio of the no. of atoms or radicals	Smallest whole no. ratio of the atoms or radical
K	8·24	39	0·2113	1
Al	5·71	27	0·2114	1
SO_4	40·49	96	0·4217	2
H_2O	45·56	18	2·531	12

Hence the empirical formula of the compound is
$$KAl(SO_4)_2 \cdot 12H_2O$$

Water molecules are written as water of crystallisation.

MOLECULAR OR TRUE CHEMICAL FORMULA

In the foregoing examples, the formulae derived for each of various substances represent their *simple* or *empirical formulae*. To determine the **True Chemical Formula** or **Molecular Formula** of a compound, we need to know not only the percentage but molecular weight of the substance as well. *The molecular formula of a substance represents the actual number of atoms of each element present in one molecule of a compound.*

RELATION BETWEEN EMPIRICAL AND MOLECULAR FORMULAE

The empirical and molecular formulae of a compound are related to each other through a simple whole number multiple. *The molecular formula of a compound is a whole number multiple of its empirical formula.*

Thus,

Molecular formula = $n \times$ Empirical formula
where n is a simple whole number integer, n may have values 1, 2, 3, 4 and so on.

The value of n is found by the simple relation,

$$n = \frac{\text{Molecular wt.}}{\text{Empirical formula wt.}}$$

Molecular weight of a substance is either given or determined from the data provided. Empirical formula weight is determined by *multiplying the atomic weight of each element with the number of atoms of the type present in the empirical formula and adding these values for all the elements in the formula.*

Example 6. *On analysis a volatile organic compound gave the following percentage composition.*

$$C = 40.00\%, H = 6.70\%, O = 53.30.$$

The V.D. of the compound is 30. What is its molecular formula?

Empirical and Molecular Formulae

Step I

Find the empirical formula as usual.

Element	At. wt. of the element	Percentage of the element	Ratio of the no. of atoms	Smallest whole no. ratio
C	12	40	$\frac{40}{12} = 3.23$	$\frac{3.33}{3.33} = 1$
H	1	6.70	$\frac{6.70}{1} = 6.70$	$\frac{6.70}{3.33} = 2$
O	16	53.30	$\frac{53.30}{16} = 3.33$	$\frac{3.33}{3.33} = 1$

Empirical formula = CH_2O

Step II

Find the empirical formula weight.

$$\text{One atom of carbon} = 1 \times 12 = 12$$
$$\text{Two atoms of hydrogen} = 2 \times 1 = 2$$
$$\text{One atom of oxygen} = 1 \times 16 = 16$$

E.F. wt. (Empirical formula wt.) = 30.

Step III

Find the molecular weight.

We know that Mol. wt. = $2 \times$ V.D.

∴ Mol. wt. of the compound = 2×30
= 60.

Step IV

Find the value of n.

$$n = \frac{\text{Mol. wt.}}{\text{E.F. wt.}}$$
$$= \frac{60}{30} = 2$$

STEP V

Find the molecular formula.

Molecular formula = $n \times$ Empirical formula

Thus molecular formula of the compound is
$$2 \times CH_2O = C_2H_4O_2$$

Example 7. *One litre of a gaseous compound at $0°C$ and 760 mm pressure weighed 4·416 g. The percentage composition of this compound is $C = 12·13\%$, $O = 61·17\%$, $Cl = 71·6·\%$. What is the molecular formula of the compound ?*

STEP I

Find the empirical formula of the compound.

Element	Percentage composition	At. wt. of the element	Ratio of the no. of atoms	Simplest whole no. ratio
C	12·13	12	1·01	$\frac{1·01}{1·01} = 1$
O	16·17	16	1·01	$\frac{1·01}{1·01} = 1$
Cl	71·69	35·5	2·02	$\frac{2·02}{1·01} = 2$

Empirical formula = $COCl_2$

STEP II

Find the Empirical formula weight.

$$\text{E.F. wt.} = 1 \times 12 + 1 \times 16 + 2 \times 35·5$$
$$= 12 + 16 + 71 = 99$$

STEP III

Find the molecular weight from the given data.

We are given that :

1 litre of the substance at 0°C and 760 mm press. (S.T.P.)

weighs = 4·416 g

Empirical and Molecular Formulae

22.4 litres of the substance at 0°C 760 mm press. (S.T.P.)

$$\text{will weigh} = 4.416 \times 22.4$$
$$= 98.9184 \text{ g}$$

Since the weight of 22.4 litres of a gas at S.T.P. is its molecular weight (G.M.W.) in grams, 99.9184 is its mol. wt.

STEP IV

Find the value of n.

$$n = \frac{\text{Mol. wt.}}{\text{E.F. wt.}} = \frac{99}{98.9184} = 1 \text{ approx.}$$

STEP V

Find the molecular formula.

Molecular formula $= 1 \times COCl_2 = COCl_2$

Example 8. *Hydrofluoric acid has the following percentage composition:* $H = 5.04\%$, $F = 95.76\%$.

Its molecule is 19.15 times heavier than the molecule of hydrogen. What is the molecular formula of hydrofluoric acid?

STEP I

Find the empirical formula of this compound.

Element	Percentage	At. wt. of the element	Ratio of the number of atoms	Simplest whole no. ratio
H	5.04	1	5.00	$\frac{5.04}{5.04} = 1$
F	95.76	19	5.04	$\frac{5.04}{5.04} = 1$

Empirical formula $= HF$

STEP II

Find the E.F. wt.

$$\text{E F wt.} = 1 \times 1 + 1 \times 19$$
$$= 1 + 19 = 20$$

Step III

Find the molecular weight.

Since the molecule of hydrofluoric acid is 19·85 times heavier than the molecule of hydrogen, its molecular weight by definitions is $19·85 \times 2$ (2 is Mol. wt. of Hydrogen).

Thus, Mol. wt. of hydrofluoric acid $= 39·70$.

Step IV

Find the value of n.

$$n = \frac{\text{Mol. wt.}}{\text{E.F. wt.}} = \frac{39}{19·85} = 2 \text{ approx.}$$

Step V

Find the molecular formula.

Molecular formula $= 2 \times HF$
$= H_2F_2$.

Example 9. *An organic compound on analysis gave the following results:*

$C = 65\%; H = 3·5\%; N = 9·6\%; O = 21·9\%.$

Its molecular weight was found to be 147. What is the molecular formula of the substance? (Punjab Pre-Univ., 1962)

Step I

Find the Empirical formula of the compound.

Element	Percentage	At. wt.	Ratio of the no. of atoms	Simplest whole no. ratio
C	65·05	12	$\frac{65}{12} = 5·41$	$\frac{5·41}{0·685} = 8$
H	3·5	1	$\frac{3·5}{1} = 3·5$	$\frac{3·5}{0·685} = 5$
N	9·6	14	$\frac{9·6}{14} = 0·685$	$\frac{0·685}{0·685} = 1$
O	21·9	16	$\frac{21·9}{16} = 1·368$	$\frac{1·308}{0·685} = 2$

Empirical formula $= C_8H_5NO_2$

Empirical and Molecular Formulae

STEP II

Find the E.F. wt.

$$\text{E.F. wt.} = 8 \times 12 + 5 \times 1 + 1 \times 14 + 2 \times 16$$
$$= 96 + 5 + 14 + 32$$
$$= 147.$$

STEP III

Find the value of n.

$$n = \frac{\text{Mol. wt.}}{\text{E.F. wt.}}$$
$$= \frac{147}{147} = 1$$

STEP IV

Find the molecular formula.

$$\text{Molecular formula} = 1 \times C_8H_5NO_2$$
$$= C_8H_5NO_2$$

END-OF-CHAPTER PROBLEMS

Empirical Formula

1. A compound contains

   $C = 52.17\%$
   $H = 13.03\%$
   $O = 34.79\%$

   Find the Empirical Formula. **[Ans. C_2H_6O]**

2. A compound contains

   $Na = 29.08\%$
   $S = 40.56\%$
   $O = 30.35\%$

   Determine then Empirical Formula **[Ans. $Na_2S_2O_3$]**

3. An oxide of manganese has

   $Mn = 63.2$
   $O = 16.8$

   Determine the Empirical Formula. **[Ans. MnO]**

4. Tin forms two oxides with following composition :

   (a) $Sn = 78.765\%$ (b) $Sn = 88.121\%$
       $O = 21.235\%$ $O = 11.879\%$.

   Determine the Empirical Formulas. **[Ans. (a) SnO (b) SnO_2]**

5. Find the empirical formula of the substance which has the following composition :

$K = 40.226\%$; $\quad$ Cr $= 26.77\%$; $\quad$ O $= 32.95\%$.

[**Ans.** K_2CrO_4]

6. Derive the empirical formula of the substance which has the following composition :

$Ca = 31.285\%$; $\quad$ C $= 18.746\%$; $\quad$ O $= 49.969\%$

[**Ans.** CaC_2O_4]

7. Derive the empirical formula of the substance which has the following composition :

(a) Fe $= 36.761\%$
$\quad$ C $= 21.106\%$
$\quad$ O $= 42.182\%$.

(b) Sb $= 83.356\%$
$\quad$ O $= 16.634\%$.

[**Ans.** $a = FeSO_4$; $b = Sb_2O_3$]

8. A salt containing water of crystallisation has the following composition :

Mg $= 9.76\%$; $\quad$ S $= 13.01\%$; $\quad$ O $= 26.01\%$; and $H_2O = 51.2\%$.

Determine the empirical formula of the compound.

[**Ans.** $MgSO_4.7H_2O$]

9. A hydrated sample of ferrous sulphate has the following composition :

Fe $= 20.15\%$; $\quad$ S $= 11.5\%$; $\quad$ O $= 23.02\%$

and water $= 45.4\%$.

Determine the empirical formula of the compound.

[**Ans.** $FeSO_4.7H_2O$]

10. A compound of Arsenic and Sulphur has As $= 69.92\%$ and S $= 39.08\%$. What is the empirical formula of the compound ?

[**Ans.** As_2S_3]

11. A compound has the following composition :

Ca $= 31.54\%$; $\quad$ P $= 24.38\%$; $\quad$ O $= 44.08\%$.

Find the empirical formula of the substance.

[**Ans.** $Ca_2P_2O_7$]

12. Analysis showed that a compound has C, H, N and O only. 1.279 g of this compound was burnt completely, 1.60 g of CO_2 and 0.77 g. H_2O were obtained. In another experiment 0.8125 g of the substance contained 0.108 g of nitrogen. What is the empirical formula of the compound ?

[**Ans.** $C_3H_8O_3N$]

13. Calculate the empirical formula of the substance having following percentage composition :

(a) Oxygen $\quad = 38.10$
(b) Hydrogen $\quad = 0.80$
(c) Phosphorus $\quad = 24.6$
(d) Sodium $\quad = 36.5$.

Empirical and Molecular Formulae 55

Name the above compound.

(At. wt. of H = 1, O = 16, Na = 23 and P = 31)

[**Ans.** Na_2HPO_3]

14. 100 g of a sample of a compound on analysis gave the following results ; 31·85 g of potassium. 28·97 g of chlorine and 39·18 g of oxygen. From this data, calculate :

(a) the number of moles of each element present in the sample.

(b) the number of atoms of each element present in the sample.

(c) the simplest ratio of the different types of atoms present in the sample.

(d) the empirical formula of the compound.

(Guru Nank Dev Pre-Univ., 1978)

[**Ans.** (a) K = 0·817 mole, Cl = 0·816, mole O = 2·45 moles

(b) K = $4·921 \times 10^{23}$ atoms

Cl = $4·915 \times 10^{23}$ atoms

O = $14·756 \times 10^{23}$ atoms

(c) K : Cl : O = 1 : 1 : 3

(d) $KClO_3$]

Molecular Formula

15. A crystalline salt on being rendered anhydrous loses 45·6% of its weight. The percentage composition of anhydrous salt is Al = 10·50%, K = 15·10%, S = 24·80% and O = 49·60. What are empirical formulae of anhydrous and hydrated salt ?

(Punjab H.S., 1975)

[**Ans.** $AlKS_2O_8$, $AlKS_2O_8.12H_2O$]

16. An oxide of nitrogen contains 2·0092 g of nitrogen combined with 5·7407 g of oxygen. The molecular weight of this oxide is 118. Find the molecular formula. [**Ans.** N_2O_5]

17. Find the molecular formulae of the compounds having the following composition :

(a) Ca = 31·29%   (b) Ca = 40·04%

   C = 18·74%   C = 11·995%

   O = 49·97%   O = 47·964%

The molecular weight of (a) is 128·07 and that of (b) is 100 07.

[**Ans.** (a) CaC_2O_3; (b) $CaCO_3$]

18. The percentage composition of an organic acid is :

C = 39·96%,   H = 6·71%,   O = 53·3%.

Its molecule is 29·78 times as heavy as hydrogen molecule. What is the molecular formula of the compound ?   [**Ans.** $C_2H_4O_2$]

19. The percentage composition of a volatile commound is :

C = 46·15%,   N = 53·85%

470 ml of the vapours of this substance weigh 1·092 g. Find the molecular formula of the substance. [**Ans.** C_2N_2]

20. A compound of carbon and hydrogen contains C = 92·23%, H = 7·77%. The V.D. of the compound is 38·2. Find the molecular formula of the compound. [**Ans.** C_6H_6]

21. On analysis a substance was found to have the following composition :

$$K = 31·84\% \qquad Cl = 28·97\%, \qquad O = 39·18\%.$$

Given that the molecular weight of the compound is 122·5. Find the molecular formula. [**Ans.** $KClO_3$]

22. An organic compound contains 65·73% Carbon, 15·06% Hydrogen, and 19·21 × Nitrogen. Its vapour density is 37. Find the molecular formula. [**Ans.** $C_4H_{11}N$]

23. Two organic compounds have the following composition : C = 35·61, H = 14·39%. The molecule of the simplest of these is 14 times heavier than the hydrogen molecule. The molecule of the second is twice as great as the first. Derive the molecular formula of each.

[First compound C_2H_4 ; second compound C_4H_8]

24. An organic compound has the following percentage composition : C = 40%, H = 6·67%, Oxygen by difference. Its V.D. is 15. Find the empirical and molecular formula of the compound.

[**Ans.** CH_2O]

25. Determine the molecular formula of a compound which contains : H = 2·1%, C = 12·8% and Br = 85·1%. 1·0 gm of it occupies 119 cc at N.T.P. (*Kurukshetra Pre-Univ., 1974*)
[**Ans.** $C_2H_4Br_2$]

26. A compound on analysis gave the following percentage composition :

$$Na = 18·59\%, \qquad S = 25·80\%, \qquad H = 4·03\%$$

and O = 51·68%. Calculate the molecular formula of the crystalline salt on assumption that all the hydrogen in the compound is present in combination with oxygen as water of crystallisation. Molecular weight of the compound is 248. (*Guru Nanak Dev Pre-Univ., 1975*)
[**Ans.** $Na_2S_2O_3 \cdot 5H_2O$]

27. A compound on analysis gave the following percentage composition :

$$Na = 14·31\%; \qquad S = 9·97\%; \qquad H = 6·22\%;$$
$$O = 69·5\%.$$

Assuming that all hydrogen in the compound is present in combination with oxygen as water of crystallisation, calculate the molecular formula of the compound. Molecular weight of the compound is 322.

(At. wt. Na = 23; S = 32; H = 1; O = 16)

(*Punjab Board Hr. Sec., 1977*)
[**Ans.** $Na_2SO_4 \cdot 10 H_2O$]

Empirical and Molecular Formulae

28. Find the molecular formula of a compound having 6·7% hydrogen, 40% carbon and rest oxygen. 0·6 g of this compound occupy 224 ml of air at S.T.P.

(Panjab Pre-Univ., 1978)
[**Ans.** $C_2H_4O_2$]

29. An organic compound on analysis gave the following results : Carbon = 65%; Hydrogen = 3·5%; Nitrogen = 9·6% and Oxygen = 21·9%, 0·3675 g of the substance displaced in a Victor Meyer's apparatus 56 ml of dry air at N.T.P. What is the molecular formula of the substance.

(Punjabi Pre-Uni., 1978 ; Punjab Board Hr. Sec., 1978)
[**Ans.** $C_8H_5NO_2$]

30. A chemical compound is found to have following composition on analysis ;

C = 19·5%; N = 22·8%; Fe = 15·2%; K = 42·5%

calculate the Empirical Formula of the compound. What will be its molecular formula, if its molecular weight is 368 ?

(At. wts. C = 12, N = 14, Fe = 56, K = 39)

(Punjab Board Hr. Sec., 1979)
[**Ans.** Empirical Formula = $C_6N_6FeK_4$
Molecular Formula = $C_6N_6FeK_4$]

31. A crystalline salt on being rendered anhydrous losses 46·4% of its weight. The percentage composition of the anhydrous salt is Al = 10·5%, K = 15·1%, S = 24·8%, O = 49·6%. Find the simplest formula of the anhydrous and the crystalline salt. (At. Wts. Al = 27, K = 39, S = 32, O = 16). *(Guru Nanak Dev Pre-Univ., 1911)*
[**Ans.** $AlKS_2O_8..12H_2O$]

32. A chemical compound on analysis gave the following percentage composition : C = 27·27%; O = 36·36%, N = 31·82%; H = 4·55%. Calculate its molecular formula if the vapour density is 44.

(At. Wts. C = 12; O = 16; N = 14; H = 1).

(Punjab School Education Board XI, 1981)
[**Ans.** $(CONH)_2$]

6

Gas Laws

The behaviour of gases and vapours is very important to a chemist. On account of their compressibility and thermal expansion, they are all affected by the changes in temperature and pressure. In general, *the volume of gases or vapours increases with increase in temperature and decrease in pressure ; and it decreases with the decrease in temperature and increase in pressure*. Their volumes are affected by these changes in temperature and pressure almost to the same extent.

GAS LAWS

All gases have been found to obey certain laws which are termed as gas laws. These gas laws relate the volume changes in a gas with temperature and pressure.

Boyle's Law. The relation between the volume of a gas at constant temperature and the pressure to which it is subjected is given in Boyle's Law. The law states that —

At a constant temperature the volume of a gas varies inversely as the pressure to which the gas is subjected.

For example, if the pressure of a given gas is doubled, the volume will be halved ; pressure is halved, the volume will be doubled, as shown in Fig. 6·1.

Gas Laws

Fig. 6·1. An illustration of Boyle's law. Temperature is constant.

Mathematically, when temperature T of the gas is kept constant, its volume V is inversely proportional to the pressure P to which it is subjected *i.e.*,

$$V \propto \frac{1}{P} \text{ at constant temperature}$$

or
$$P \propto \frac{1}{V}$$

or
$$P \times V \propto K \text{ (a constant)}$$

or
$$PV = \text{constant}.$$

If we change the pressure of the gas from a volume P to P_1, the volume will also change. Let the changed volume be V_1.

According to the law even $P_1 \times V_1$ will also be constant and will have the same value as $P \times V$

Numerical Problems in Chemistry

So that $P \times V = P_1 \times V_1$

i.e., initial pressure × initial volume = changed pressure × changed volume.

The law may also be stated as "*At constant temperature the product of pressure and volume of a gas is a constant quantity*"

Example 1. *A certain mass of 2·5 litres of oxygen has a pressure of 740 mm. What volume will the same mass of oxygen have under a pressure of 780 mm ?*

In this question :

P (initial pressure) = 740 mm

V (initial volume) = 2·5 litres

P_1 (changed pressure) = 780 mm

V_1 (changed volume) = ?

Applying Boyle's law,

$$P \times V = P_1 \times V_1$$

and substituting the value of P_1, V_1, P_2, we have

$$740 \times 2·5 = 780 \times V_1$$

or $V_1 = \dfrac{740 \times 2·5}{780} = 2·371$ litres.

Example 2. *A gas was enclosed in a cylinder which has a movable piston. The volume of the gas in the cylinder was 500 ml and the pressure was 1 atmosphere (760 mm of Hg). What is the resulting pressure of the gas when the volume increases to 1000 ml ?*

Initial volume of the gas (V) = 500 ml

Initial pressure of the gas (P) = 760 ml

Changed volume (V_1) = 1000 ml (1 litre)

Changed pressure (P_1) = ?

Applying Boyle's law

$$P \times V = P_1 \times V_1$$

and substituting the values of P, V and V_1, we have

$$760 \times 500 = P_1 \times 1000$$

or $P_1 = \dfrac{760 \times 500}{1000} = 380$ mm of Hg

Example 3. *It is desired to pump 50 litres of air which was measured at 720 mm pressure into an 8-litre vessel. What pressure will the air exert in the new vessel ?*

It is given in this problem that—
Initial pressure P = 720 mm
,, volume V = 50 litres
Now volume V_1 = 8 litres
New pressure P_1 = ?

By substituting the values in $P \times V = P_1 \times V_1$, we have

$$720 \times 50 = P_1 \times 8$$

$$P_1 = \frac{720 \times 50}{8} = 4500 \text{ mm}$$

CHARLES LAW

In 1787 Charles described the variation of the volume of a gas with respect to the changes in the temperature of its environments. It was noticed that when the temperature of

Fig. 6·2. An illustration of Charles' law (temperature in degrees Kelvin). Pressure is constant.

a gas was increased, its volume also increased provided, of course, the pressure was kept constant. Charles' law states:

"Provided that the pressure is kept constant, the volume of a given mass of gas will vary directly with the absolute temperature."

For example, if the Kelvin temperature is doubled at constant pressure, the volume is doubled, and if the Kelvin temperature is halved, the volume is halved, as shown in Fig. 6.2.

This is mathematically expressed as

$$V \propto T \quad \text{when P is kept constant}$$

or
$$\frac{V_1}{T_1} = \frac{V_2}{T_2}$$

or
$$\frac{V_1}{V_2} = \frac{T_1}{T_2}$$

where V_1 and T_1 are the initial volume and absolute temperature and V_2 and T_2 are the final volume and absoulte temperature respectively.

Example 4. *When measured at 27°C, the volume of oxygen was 400 ml. What would be the new volume at 102°C ? (Pressure remaining constant)*

It is given that

$$V_1 = 400 \text{ ml}$$
$$T_1 = 273 + 27 = 300°K \text{ (or abs.)}$$
$$V_2 = ?$$
$$T_2 = 273 + 102 = 375°K$$

Substituting the values in the equation,

$$\frac{V_1}{T_1} = \frac{V_2}{T_2} \text{ we have}$$

$$\frac{400}{300} = \frac{V_2}{375}$$

$$V_2 = \frac{400 \times 375}{300} = 500 \text{ ml}$$

Example 5. *Given 400 ml of a gas at 17°C, what temperature must be attained in order to (a) double the volume (b) reduce the volume by one half.*

Gas Laws

It is given that
$$V_1 = 400 \text{ ml}$$
$$T_1 = 273 + 17 = 290°K \text{ (abs.)}$$
$$T_2 = ?$$
$$V_2 = 2 \times 400 \text{ (in part } a) \text{ and } \tfrac{1}{2} \times 400 \text{ in } (b)$$

(a) Substituting the values in the expression
$$\frac{V_1}{T_1} = \frac{V_2}{T_2}$$
$$\frac{400}{290} = \frac{800}{T_2}$$
$$T_2 = \frac{800 \times 290}{400} = 580°K$$
$$= 580 - 273 = 307°C$$

(b) Substituting the values in
$$\frac{V_1}{T_1} = \frac{V_2}{T_2}$$
$$\frac{400}{290} = \frac{200}{T_2}$$

or
$$T_2 = \frac{200 \times 290}{400} = 145°K$$
$$= 145 - 273 = -128°C$$

Gas Equation. The above two gas laws can be employed to derive the general gas law or the gas equation which applies to the changes of temperature and pressure on the volume of a gas. The gas equation is

$$\frac{P_1 V_1}{T_1} = \frac{P_2 V_2}{T_2} = \text{constant for a given mass of a gas.}$$

Example 6. *Given 200 ml of a gas at 5°C and 760 mm. Find the volume at 30°C and 800 mm.*

It is given that
$$P_1 = 760 \text{ mm}$$
$$V_1 = 200 \text{ ml}$$
$$T_1 = (5+273)° \text{ Abs. } 278° \text{ Abs.}$$
$$P_2 = 800 \text{ mm}$$
$$V_2 = ?$$
$$T_2 = (273+30)° \text{ Abs. } = 303° \text{ Abs.}$$

Substituting the values in the gas equation

$$\frac{P_1 V_1}{T_1} = \frac{P_2 V_2}{T_2}, \text{ we have}$$

$$\frac{760 \times 200}{278} = \frac{800 \times V_2}{303}$$

or
$$V_2 = \frac{760 \times 200 \times 303}{800 \times 278}$$

$$= 207 \cdot 8 \text{ ml.}$$

Example 7. *A mass of a gas occupies 0·084 litre at 300°C and 25 atmospheres. What volume will it occupy at S.T.P. ?*

It is given that

$V_1 = 0.084$ litre
$P_1 = 25$ atm.
$T_1 = (273+300)°$ Abs. $= 573°$ Abs.
$P_2 = 1$ atmosphere
$V_2 = ?$
$T_2 = 273°$ Abs.

Substituting the values in the gas equation

$$\frac{P_1 V_1}{T_1} = \frac{P_2 V_2}{T_2}, \text{ we have}$$

$$\frac{25 \times 0.084}{573} = \frac{1 \times V_2}{273}$$

or
$$V_2 = \frac{25 \times 0.084 \times 273}{573 \times 1}$$

$$= \frac{573 \cdot 3}{573} = 1 \text{ litre.}$$

Dalton's Law of Partial Pressures. When two gases occupy the same container each gas behaves as if it alone occupied that container and exerts its own individual pressure. Naturally, the total pressure exerted by the mixture is equal to sum of the individual pressures exerted by various gases present in the mixture. This generalisation has been stated in the form of **Dalton's Law of Partial Pressures.** The law states that :

"The total pressure exerted by a mixture of gases is equal to sum of all the partial (or individual) pressures that each gas would exert if present alone in that space."

Gas Laws

In the atmosphere a variety of gases are present. The atmospheric pressure is due to the sum of the individual pressures exerted by all those gases that constitute the atmosphere. This may be expressed as,

$$P_{atmosphere} = P_{oxygen} + P_{nitrogen} + P_{carbon\ dioxide} + P_{inert\ gases} + etc.$$

Similarly, when a gas is collected over a volatile liquid, such as water, the gas takes away with it some vapours of this liquid (like water vapours). This gas, which is now saturated with water vapours and occupies certain volume, exerts some pressure. This pressure actually is the sum total of two partial pressures *(i)* Pressure of the gas itself *(ii)* Pressure of water vapours.

Thus we write that :

Pressure of the moist gas ($P_{moist\ gas}$)
= Pressure of the dry gas (P_{dry})
+ Pressure of water vapour ($P_{moisture}$)

or $\quad P_{moist\ gas} = P_{dry\ gas} + P_{moisture}$

The pressure of the water vapours is also called *Aqueous tension*.

Thus,

$$P_{moist\ gas} = P_{dry\ gas} + \text{Aqueous Tension (Aq. tension)}$$

or $\quad P_{dry\ gas} = P_{moist\ gas} - P_{moisture}$ (or Aq. tension)

Thus when a gas is collected over water and has certain pressure, to find the pressure of the dry gas we have to subtract the Aq. tension from the total pressure. Aq. tension has a definite value for each temperature. When, however, a gas is collected over mercury no such correction is needed as mercury has negligible pressure.

Example 8. *2 litres of oxygen, 1 litre of nitrogen and 2 litres of CO_2 measured at S.T.P. were pumped into a container having 1 litre capacity. What is the total pressure exerted by these gases in the new container under similar condition of temperature ?*

STEP I

Calculate the pressure exerted by each gas if it alone occupied the container.

Now for oxygen, the initial volume is 2 litres but the final volume is one litre. Since the temperature is identical in the beginning and the end, the pressure will be doubled according to Boyle's Law.

i.e.
$$P_1V_1 = P_2V_2$$
$$P_1 = 1 \text{ atm}$$
$$V_1 = 2 \text{ litres}$$
$$P_2 = ?$$
$$V_2 = 1 \text{ litre}$$

$$\therefore P_2 = \frac{P_1V_1}{V_2} = \frac{1 \times 2}{1} = 2 \text{ atm}$$

For nitrogen, since the intial volume, pressure and temperature are similar to final volume, pressure and temperature, it will exert the same pressure *i.e.*, 1 atmosphere.

For carbon dioxide like oxygen, the initial and final temperatures are same. The initial volume (2 litres) is half of the final volume (1 litre), the pressure will be doubled (according to Boyle's law).

Thus the pressure of $CO_2 = 2$ atmospheres at the final stage.

STEP II

Apply the partial pressure equation.

Total pressure of the mixture $(P_{total}) = P_{O_2} + P_{N_2} + P_{CO_2}$
$$= 2 \text{ atm} + 1 \text{ atm} + 2 \text{ atm}$$
$$= 5 \text{ atmospheres}$$
$$= 5 \times 760 = 3800 \text{ mm}$$

Example 9. *A mixture of 40% N_2, 40% O_2 and 20% CO_2 by volume was enclosed in a space. The total pressure of the gas mixture was 800 mm. What is the partial pressure of each gas ?*

Each component of the mixture occupies the entire volume of the mixture. We have actually to find the percentage composition by volume of each gas before mixing.

Now 100 volumes of the mixture which contains 40 volumes of nitrogen, 40 vols of oxygen and 20 vols of carbon dioxide exert a total pressure of 800 mm. We shall now divide this pressure in the ratio of their respective volumes.

Thus the partial pressure of N_2 = $\dfrac{800}{100} \times 40$

= 320 mm

the partial pressure of O_2 = $\dfrac{800}{100} \times 40$

= 320 mm

the partial pressure of CO_2 = $\dfrac{800}{100} \times 20$

= 160 mm.

Example 10. *A sample of oxygen collected over water occupied a volume of 325 ml at 25°C and exerted a pressure of 720 mm of Hg. Find the pressure of the dry gas. Also find the volume of oxygen at S.T.P. (Aq. tension at 25°C is 23·7 mm)*

STEP I

Find the volume of the dry gas.

According to Dalton's law of partial pressure

$P_{moist\ gas}$ = $P_{dry\ gas} + P_{moisture}$
$P_{dry\ gas}$ = $P_{moist\ gas} - P_{moisture}$ (or Aq. tension)
= 720 − 23·7
= 696·3 mm

∴ P_1 = 696·3 mm
V_1 = 325 ml
T_1 = 273 + 25
= 298° Abs

STEP II

Find the vol. of oxygen at S.T.P.

Applying the gas equation and substituting the values of P_1, V_1, T_1, P_2, T_2 in the equation $\left(\dfrac{P_1 V_1}{T_1} = \dfrac{P_2 V_2}{T_2} \right)$ we have

$$\dfrac{696 \cdot 3 \times 325}{298} = \dfrac{760 \times V_2}{273}$$

$$V_2 = \frac{696\cdot3 \times 325 \times 273}{298 \times 760}$$

$$= 272\cdot78 \text{ ml.}$$

END-OF-CHAPTER PROBLEMS

Boyle's Law

1. A sample of nitrogen occupied a volume of 45 ml at 740 mm pressure. What would be the volume of the same gas at 700 mm? Temperature remaining constant. **[Ans. 47·5 ml]**

2. A mass of oxygen occupies 20 c.ft. at 758 mm. What is the volume at 635 mm. Temperature remaining constant? **[Ans. 23.85 c.ft.]**

3. 40 ml of N_2 was collected at a pressure of 700 mm. What pressure must be applied to make this gas occupy 15 ml? Temperature remains constant. **[Ans. 1866·6 mm]**

4. A gas occupying 250 ml and under a pressure of 745·5 mm Hg was allowed to occupy a volume 437.5 ml the temperature remaining constant. What will be the resulting pressure of the gas? **[Ans. 426 ml]**

5. If 1.78 litres of oxygen were measured at 60 mm, what volume would the gas occupy at 760 mm? The temperature remained constant throughout. **[Ans. 1·405 litres]**

6. What pressure must be applied to a given sample of a gas in order to compress it to (a) one-fourth; (b) one-third; (c) two-third of its original volume?

  [Ans. (a) four times the original pressure;

  (b) three times the original pressure;

  (c) $\frac{3}{2}$ times the original pressure]

7. The volume of a gas with some marble measures 100 c.c. at 760 mm pressure. On raising the pressure to 1000 mm the total volume becomes 80 c.c. What is the volume occupied by the marble? *(Dibrugarh Pre-Univ. 1973)* **[Ans. 16·67 c.c.]**

8. A given mass of a gas occupies 200 c.c. at a pressure of 760 mm. What volume will it occupy at 780 mm pressure while keeping the temperature constant? *(Aligarh Pre-Univ., 1973)* **[Ans. 194·8 c.c.]**

9. 200 c.c. of Nitrogen at 17°C are cooled to −20°C. Find the new volume of the gas at the same pressure. *(Punjabi Pre-Univ., 1973)* **[Ans. 174·48 c.c.]**

Charles' Law

10. A given mass of oxygen occupies 76 ml at 20°C. Determine its volume at 45°C, pressure remaining constant. **[Ans. 82·4 ml]**

11. A sample of chlorine has a volume of 890 ml at 233°C. What will be its volume at 3°C, pressure remaining constant?

[**Ans.** 480 ml]

12. Given 200 ml of a gas at 18°C. What temperature must be attained in order to (a) double the volume of the gas (b) reduce it to one half, pressure remaining constant?

[**Ans.** $a = 309°C$;
$b = -127.5°C$]

13. Calculate the volume of 690 ml of a gas at 21°C when cooled to −10°C, pressure remaining constant. [**Ans.** 618 ml]

14. What volume 20.0 cubic feet of nitrogen measured at 12°C, occupy at 36°C, pressure remaining constant? [**Ans.** 21.6 cu. ft.]

15. The volume of a sample of gas, when measured at 17°C was 9.0 cu.ft. To what temperature should the gas be heated in order that its volume shall be (a) doubled (b) tripled; pressure remaining constant? [**Ans.** $a = 307°C$; $b = 597°C$]

16. How much volume does 30 ml of hydrogen collected at 30°C and atmospheric pressure occupy at N.T.P?

(*Sambalpur Pre-Univ., 1973*)
[**Ans.** 27.04 ml]

17. At what temperature will a given volume of a gas at 0°C double itself, pressure remaining constant?

(*Himachal Pre-Univ., 1977*)
[**Ans.** 273°C]

18. 400 ml of oxygen at 27°C were cooled to −15°C without the change in pressure. Calculate the contraction in volume.

(*Haryana Board Hr. Sec., 1979*)
[**Ans.** 56 ml]

Gas Equation

19. What volume will 1000 ml of oxygen at 710 mm and 17°C occupy under a pressure of 550 mm and a temperature of −13°C?

[**Ans.** 1157.2 ml]

20. 500 ml of oxygen at 0°C and 760 mm was subjected to a pressure of 20 atmospheres and the temperature increased to 12°C. Calculate the new volume.

[**Ans.** 26.09 ml]

21. What pressure must be applied to a gas in order that its volume may measure 53 ml at 21°C, when its volume at 18°C and 753 mm is 125 ml? [**Ans.** 1914.6 mm]

22. 250 ml of oxygen is collected over water at 25°C and 755 mm. The gas is saturated with water vapour. Find the volume of the dry gas at S.T.P. (Aq. tension at 25°C is 23.8 mm)

[**Ans.** 220.5 ml]

23. 50 ml of hydrogen were collected over water at 15°C and 770 mm pressure. What volume will the dry gas occupy at S.T.P?
(Aq. tension at 15°C is 12.7 mm) [**Ans.** 47.24 ml]

24. Calculate the temperature necessary for 125 ml of dry air at 22°C and 755 mm to a volume of 50 ml at 700 mm. [Ans. —164°C]

25. A mass of gas occupies 412·5 ml at —30°C and 422·56 mm. What is the pressure if this volume changes to 500 ml and temperature to 20°C ? [Ans. 420·28 ml]

26. Calculate the centigrade temperature required to change 5 litres of oxygen at 100°K and 0·20 atm to 10 litres at 0·4 atm.

[Ans. 127°C]

27. One litre of ammonia at N.T.P. weighs 0·76 gram. Find the weight of ammonia which measures one litre at 27°C and 750 mm pressure. *(Shivaji Pre-Univ., 1975)*

[Ans. 0·6825 g]

28. A flask of 2·5 litre capacity can withstand a pressure of 14 atmospheres and is filled with CO_2 at 27°C and 750 mm pressure. At what temperature it will explode. *(Panjab Pre-Univ., 1978)*

[Ans. 3928°C]

29. 250 ml of oxygen gas are collected over water at 25°C temperature and 750 mm pressure. Find the volume of the dry gas at N.T.P. (Aq. tension at 25°C = 23·8 mm)

(Punjab School Education Board XI 1983)

[Ans. 218·84 mm]

Dalton's Law of Partial Pressures

30. 500 ml of nitrogen and the same volume of oxygen measured separately at 760 mm pressure were led into 1 litre flask. What is resulting pressure in the flask ? (Temperature remaining constant)

[Ans. 760 mm]

31. 1000 ml capacity flask contained three gases at a total pressure of 1500 mm. The ratio of the partial pressures of the gases were 2 : 0·5 : 0·5. What volume would each gas occupy separately at 760 mm ?

(Temperature remaining constant throughout).

[**Hint** The partial pressures of the three gases are:

$$\frac{1500}{3} \times 2, \quad \frac{1500}{3} \times 0.5, \quad \frac{1500}{3} \times 0.5]$$

[Ans. The volumes at 760 will be 1st gas = 1315·78 ml ;

2nd gas = 328·94 ml :

3rd gas = 398·94 ml]

32. If 200 ml of oxygen and 720 mm : 400 ml of nitrogen at 600 mm ; and 250 ml of hydrogen at 750 mm are forced into a 1-litre flask, what will be the resulting pressure ? [Ans. 571·5 mm]

33. Air contains nitrogen 78·3%, oxygen 20·99%, rare gases 0·94% and carbon dioxide 0·04% by volume. What is the partial pressure of each gas when the atmospheric pressure is 740 mm ?

[Ans. N_2 = 577·4 mm ;

O_2 = 155·3 mm ;

rare gases = 6·95 mm ;

CO_2 = 0·296 mm ;

Gas Laws

34. A 250 ml flask was filled with 150 ml of hydrogen at 750 mm ; 75 ml of oxygen at 350 mm ; and 50 ml of carbon dioxide at 250 mm. Determine (a) the partial pressure of each gas after mixing ; (b) the total pressure of the mixture in the flask.

[**Ans.** (a) Pressure of H_2 = 450 mm ;
Pressure of O_2 = 105 mm ;
Pressure of CO_2 = 50 mm ;
(b) Total pressure = 605 mm]

35. A sample of hydrogen gas collected over water has a volume of 60 cc measured at 27°C and 596.5 mm pressure. Calculate the volume of the gas at NTP. (Aqueous tension at 27°C=26.5 mm).
(Marathwada Pre-Univ., 1976)

[**Ans.** 55.95 ml]

36. Calculate the total pressure in atmospheres in a mixture of 71 g of chlorine and 32 g of oxygen at 0°C, if the volume occupied by the mixture is 22.4 litres. *(Madurai Pre-Univ., 1976)*

[**Ans.** 2 atm]

37. A vessel of 120 cc contains a certain mass of a gas at 20°C and 750 mm pressure. The gas was transferred to a vessel whose volume is 180 cc. Calculate the pressure of the gas at 20°C.
(Kurukshetra Pre-Univ., 1976)

[**Ans.** 500 mm]

38. The volume of a gas is 200 ml at 12°C and 750 mm pressure. What volume will it occupy at 40°C and 720 mm pressure ?
(Kashmir Pre-Univ., 1976)

[**Ans.** 228.8 ml]

39. A gaseous mixture of carbon dioxide and carbon monoxide contains 25 mole per cent carbon dioxide. If the total pressure in the container is 6.0 atm, what are the partial pressures of each of the two gases ? *(Gujarat Pre-Univ., 1976)*

[**Ans.** CO_2=1.5 atm ; CO=4.5 atm]

40. The partial pressures of H_2, O_2 and N_2 in a container are 350 mm of Hg, 26.00 mm of Hg and 0.50 atm. respectively. What is the total pressure in the container ?
(Guru Nanak Dev Pre-Univ., 1977)

[**Ans.** 756 mm of Hg]

41. A container contains a mixture of three gases O_2, H_2 and N_2. The partial pressures of O_2, H_2 and N_2 are 0.2 atm of Hg, 525 mm of mercury and 17 cm of Hg respectively. What is the total pressure of gases in the container.
(Guru Nanak Dev Pre-Univ., 1978)

[**Ans.** Total pressure = 84.7 cm of Hg]

42. A gaseous mixture containing 0.025 moles of nitrogen and 16.00 gms of oxygen is contained in a flask of five litres capacity at 27°C. Calculate : (a) the partial pressure of each gas ; (b) the total pressure of gaseous mixture.
(Guru Nanak Dev Pre-Univ., 1982)

[**Ans.** (a) N_2 = 1.12 atmospheres, O_2 = 2.3 atmospheres ;
(b) 3.42 atmospheres].

7

Diffusion of Gases

When two gases contained in two different cylinders are brought in contact with each other by joining the two cylinders after removing their lids, it is found that the two gases start mixing gradually until a homogeneous mixture is produced. Also, when two cylinders containing one and the same gas but at different pressures are connected to each other, a uniform pressure will result throughout the two cylinders. **This phenomenon by virtue of which gases intermix with each other is known as diffusion**. This process of diffusion is actually a consequence of the kinetic theory of gases.

GRAHAM'S LAW OF DIFFUSION

This law governs the rate of diffusion of gases. By the rate of **diffusion** we mean the volume of the gas diffused per unit time. Mathematically,

$$\text{Rate of diffusion} = \frac{\text{Volume of the gas diffused}}{\text{Time taken}}$$

The rates of diffusion of different gases differ from each other. The lighter gas (*i.e.*, one of low density) diffuses faster than the one which is relatively heavier (*i.e.*, which has greater density). The rate of diffusion also depends upon temperature and pressure. Thus in comparing the rates of diffusion, the gases must be under similar conditions of temperature and pressure. The relation between the densities and the rate of diffusion of gases is given by **Graham's Law**. Graham's law states :

"*Under similar conditions of temperature and pressure, the rates of diffusion of gases are inversely proportional to the square root of their densities.*"

Diffusion of Gases

Mathematically,

$$\frac{r_1}{r_2} \propto \sqrt{\frac{d_2}{d_1}}$$

where r_1 and r_2 are the rates of diffusion of two gases whose densities are d_1 and d_2 respectively.

Rates of diffusion and Molecular weights. Density is directly proportional to the molecular weight of the gas. We know that mol. wt. $= 2 \times$ V.D., thus the above relation can be written as

$$\frac{r_1}{r_2} = \sqrt{\frac{2 \times d_2}{2 \times d_1}} = \sqrt{\frac{M_2}{M_1}}$$

Times of diffusion of gases. The rate of diffusion of a gas is **inversely** proportional to the time of diffusion. For instance, if a gas has a greater rate of diffusion, it will take lesser time to diffuse across a definite area of cross section. Thus

$$r \propto \frac{1}{t}$$

The Graham's Law may then be written as

$$\frac{r_1}{r_2} = \frac{t_2}{t_1} = \sqrt{\frac{d_2}{d_1}}$$

where t_1 and t_2 are the times of diffusion of the two gases whose rates of diffusion are r_1 and r_2 respectively.

Graham's Law of diffusion is helpful in finding the molecular weight, densities or times of diffusion of gases.

RELATIVE RATES OF DIFFUSION

Type I

Example 1. *Find out the relative rates of diffusion of oxygen and carbon dioxide. Molecular weight of oxygen is 32 and that of carbon dioxide is 44.*

Applying the Graham's Law and substituting the value of the molecular weights of O_2 and CO_2, we have

$$\frac{rO_2}{rCO_2} = \sqrt{\frac{M_{CO_2}}{M_{O_2}}}$$

$$= \sqrt{\frac{44}{32}} = \frac{\sqrt{1\cdot 375}}{\sqrt{1}} = \frac{1\cdot 172}{1}$$

Thus, the relative rates of diffusion of CO_2 and O_2 are

$$\frac{1\cdot 172}{1}.$$

Example 2. *Find the relative rates of diffusion of CO_2 and CO when their molecular weights are 44 and 28 respectively.*

Using the relation $\dfrac{r_{CO_2}}{r_{CO}} = \sqrt{\dfrac{M_{CO}}{M_{CO_2}}}$

and substituting the values, we have

$$\frac{r_{CO_2}}{r_{CO}} = \sqrt{\frac{28}{44}} = \sqrt{\frac{1}{1\cdot 55}}$$

Hence the relative rates of diffusion of CO_2 and CO are

$$\frac{1}{1\cdot 55}.$$

MOLECULAR WEIGHTS AND DENSITIES OF GASES

Type 2

Example 3. *The relative rate of diffusion of a gas as compared to that of CO_2 is 226 : 213. Calculate the molecular weight of the gas.*

It is given that

Mol. wt. of CO_2 = 44

Rate of diffusion of the gas = 226

,, ,, ,, ,, CO_2 = 213

Using the relation $\dfrac{r_{gas}}{r_{CO_2}} = \sqrt{\dfrac{M_{CO_2}}{M_{gas}}}$

$$= \frac{226}{213} = \sqrt{\frac{44}{M_{gas}}}$$

Squaring both sides

$$\frac{226 \times 226}{213 \times 213} = \frac{44}{M_{gas}}$$

Diffusion of Gases

or Mol. wt. of the gas (M_{gas}) $= \dfrac{44 \times 213 \times 213}{226 \times 226}$

$= 39 \cdot 08$

Example 4. *If 200 ml of oxygen take 25 seconds to diffuse through a porous pot and 400 ml of chlorine take 70 seconds under similar conditions of temperature and pressure, find the density of oxygen if that of chlorine is 35.*

It is given that

The volume of chlorine which diffuses in 70·0 seconds

$= 400$ ml

Volume of oxygen which diffuses in 25 seconds

$= 200$ ml

STEP I

Find the rates of diffusion of O_2 and Cl_2.

Rate of diffusion of $O_2 = \dfrac{200}{25} = 8$ ml/sec.

Rate of diffusion of $Cl_2 = \dfrac{400}{70} = \dfrac{40}{7}$ ml/sec.

STEP II

Find the density of chlorine.

Applying Graham's Law of diffusion and substituting the values, we have

$$\dfrac{rO_2}{rCl_2} = \sqrt{\dfrac{dCl_2}{dO_2}}$$

$$\dfrac{8 \times 7}{40} = \sqrt{\dfrac{35}{dO_2}}$$

Squaring both sides, we have

$$\dfrac{8 \times 8 \times 7 \times 7}{40 \times 40} = \dfrac{35}{dO_2}$$

or $\qquad dO_2 = \dfrac{35 \times 40 \times 40}{7 \times 7 \times 8 \times 8} = 17 \cdot 87$

Thus the relative density of oxygen is 17·87

Example 5. *The relative rates of diffusion of chlorine and ozone are 1·644 and 2. The molecular weight of chlorine is 71, find that of ozone.*

Applying the Graham's Law of diffusion and substituting the values, we have

$$\frac{rCl_2}{rO_3} = \sqrt{\frac{Mo_3}{MCl_2}}$$

$$\frac{1 \cdot 644}{2} = \sqrt{\frac{Mo_3}{71}}$$

Squaring both sides, we have

$$\frac{1 \cdot 644 \times 1 \cdot 644}{2 \times 2} = \frac{Mo_3}{71}$$

or Mol. wt. of Ozone, $(Mo_3) = \frac{1 \cdot 644 \times 1 \cdot 644 \times 71}{2 \times 2}$

$$= 47 \cdot 97$$

TIME OF DIFFUSION

TYPE 3

Example 6. *If 40·0 ml of hydrogen diffuse in 2 seconds, what time will 250 ml of chlorine take to diffuse under the same conditions of temperature and pressure? Assuming that the densities or Mol. wts. of H_2 and Cl_2 are known.*

STEP I

Find the rates of diffusion of hydrogen and chlorine.

The rate of diffusion of hydrogen

$$rH_2 = \frac{\text{Vol. of } H_2 \text{ diffused}}{\text{Time taken}}$$

$$= \frac{40}{2} = 20 \text{ ml/sec.}$$

Let the time taken by 250 ml of chlorine to diffuse
$$= x \text{ seconds}$$

$\therefore$ Rate of diffusion $Cl_2 (rCl_2) = \dfrac{250}{x}$ ml/sec.

Step II

Find the time of diffusion of Cl_2 gas.

Applying Graham's Law of diffusion and substituting the values, we have

$$\frac{r_{H_2}}{r_{Cl_2}} = \sqrt{\frac{d_{Cl_2}}{d_{H_2}}}$$

or

$$\frac{\frac{20}{250}}{x} = \sqrt{\frac{d_{Cl_2}}{d_{H_2}}}$$

$$\frac{20 \times x}{250} = \sqrt{\frac{d_{Cl_2}}{d_{H_2}}}$$

Squaring both sides

$$\frac{20 \times 20 \times x \times x}{250 \times 250} = \frac{d_{Cl_2}}{d_{H_2}} = \frac{35 \cdot 5}{1}$$

or

$$\frac{20 \times 20 \times x^2}{250 \times 250} = \frac{35 \cdot 5}{1}$$

or

$$x^2 = \frac{35 \cdot 5 \times 250 \times 250}{20 \times 20}$$

or

$$x = 74 \cdot 47$$

Thus time taken by 250 ml chlorine to diffuse
= 74·47 seconds.

Example 7. *Equal volume of hydrogen and oxygen diffused is 3 and 12 minutes respectively under similar conditions of temperature and pressure. What is the density of oxygen if that of hydrogen is unity ?*

From the Graham's Law we know that the time of diffusion are directly proportional to the quare root of the densities of the diffusing gases.

i.e.,

$$\frac{t_1}{t_2} = \sqrt{\frac{d_1}{d_2}}$$

or

$$\frac{t_{H_2}}{t_{O_2}} = \sqrt{\frac{d_{H_2}}{d_{O_2}}}$$

Substituting the values in the above relation, we have

$$\frac{3}{12} = \sqrt{\frac{d_{H_2}}{d_{O_2}}}$$

Squaring both sides

$$\frac{3 \times 3}{12 \times 12} = \frac{d_{H_2}}{d_{O_2}}$$

or

$$\frac{3 \times 3}{12 \times 12} = \frac{1}{d_{O_2}}$$

or

$$d_{O_2} = \frac{12 \times 12}{3 \times 3} = 16.$$

VOLUME OF DIFFUSING GASES

Type 5

Example 8. *Find the volume of oxygen which takes 100 seconds to diffuse when 20 ml of CO_2 take 46·8 seconds to diffuse under similar conditions of temperature and pressure. Assuming the densities of O_2 and CO_2 to be known.*

STEP I

Find the rates of diffusion of O_2 and CO_2.

Let the volume of O_2 which diffuses in 100 sec. $= x$ ml.

∴ Rate of diffusion of O_2 (r_{O_2}) $= \dfrac{x}{100}$ ml/sec.

Similarly the rate of diffusion of CO_2 (r_{CO_2})

$$= \frac{20}{46\cdot 8} \text{ ml/sec.}$$

STEP II

Find the value of x.

Applying Graham's Law of diffusion and substituting the values, we have

$$\frac{r_{O_2}}{r_{CO_2}} = \sqrt{\frac{d_{CO_2}}{d_{O_2}}}$$

Diffusion of Gases

$$\frac{\dfrac{x}{100}}{\dfrac{20}{46 \cdot 8}} = \sqrt{\frac{44}{32}}$$

Squaring both sides, we have

$$\frac{46 \cdot 8 \times 46 \cdot 8 \times x \times x}{100 \times 100 \times 20 \times 20} = \frac{44}{32}$$

or
$$x^2 = \frac{44 \times 100 \times 100 \times 20 \times 20}{46 \cdot 8 \times 46 \cdot 8 \times 32}$$

or
$$x^2 = 2512 \cdot 0 \quad \text{or} \quad x = \sqrt{2512} = 50 \cdot 12$$

Thus the volume of oxygen which diffuses in 100 seconds is 50·12 ml.

END-OF-CHAPTER PROBLEMS

Densities and Molecular Weights

1. The speeds of diffusion of a gas relative to CO_2 are 9 : 9·66. Calculate the relative density of the gas when that of CO_2 is 22.
[Ans. 25·3]

2. Hydrogen diffuses 4 times as rapidly as a certain gas. What is the density and molecular weight of the gas when that of Hydrogen is unity ?
[Ans. 16]

3. In 25 seconds 150 ml of oxygen diffuses through a porous pot while 500 ml of another gas diffuses in 124·2 seconds. Assuming the density of oxygen to be known, find that of the other gas. Also find its molecular weight.
[Ans. 35·5 ; 71]

4. Calculate the molecular weight of nitrogen when the relative rates of diffusion of N_2 and O_2 are 1·07 : 1. The molecular weight of O_2 being known.
[Ans. 28]

5. Calculate the molecular weight of aqueous vapours when the relative rates of diffusion of hydrogen and the aqueous vapours are 2 : 0·335. The density of hydrogen being 1.
[Ans. 18]

6. 84·5 ml of oxygen diffuses in the same time as 60 ml of sulphur dioxide. Find the density of sulphur dioxide when that of oxygen is 16.
[Ans. 32]

7. 40 ml of hydrogen take 8 seconds to diffuse through a porous pot while 20 ml of nitrogen diffuses in 14·95 seconds under similar conditions. If the density of hydrogen is unity, find that of nitrogen.
[Ans. 13·95]

8. In an experiment it was found that 57 ml of hydrogen take the same time to diffuse through a porous partition as 10 ml of sulphur dioxide under similar conditions. Calculate the molecular weight of sulphur dioxide.
[Ans. 64·9]

9. Under the same conditions of temperature and pressure a given volume of hydrogen escaped through a fine hole in 26 seconds whereas equal volume of carbon dioxide required 122 seconds. Find the density of CO_2 relative to hydrogen.
[Ans. 22·02]

10. An unknown gas diffuses four times as quickly as oxygen. Calculate the molecular weight of the gas. **[Ans. 2]**

11. The relative rate of diffusion of a gas as compared with carbon dioxide is 27 : 29. Calculate the molecular weight of the gas. The vapour density of carbon dioxide is 22. *(Punjab H.S. 1974)* **[Ans. 50·77]**

12. Two gases CO_2 and an unknown gas having same volume diffuse through a porous pot in 14 and 10 seconds respectively. Calculate the molecular weight of unknown gas. *(Himachal Pre-Univ., 1976)* **[Ans. 22·44]**

13. 127 cc of a certain gas diffuse in the same time as 100 cc of chlorine under same conditions of temperature and pressure. Calculate the molecular weight of the gas. *(Punjabi Pre-Univ., 1978)* **[Ans. 44·02]**

14. Two gases A and B having the same volume diffuse through a porous partition in 20 and 10 seconds respectively. The molecular weight of A is 49. Calculate the molecular weight of B.
(Kurukshetra Pre-Univ., 1982); *(Himachal Pre-Univ., 1981)* **[Ans. 196]**

15. Two gases CO_2 and unknown gas having same volume diffuse through a porous partition in 14 and 10 seconds. Calculate the molecular weight of unknown gas. (At. Wts. C—12. O = 16).
(Maharshi Dayanand Pre-Univ., 1982) **[Ans. 86·24]**

16. 112 ml of hydrogen diffuse in the same time as 28 ml. of an unknown gas. Find the molecular weight of the gas, if that of hydrogen be 2. *(Punjabi Pre-Univ., 1982)* **[Ans. 64]**

Time of Diffusion

17. In an experiment 100 ml of hydrogen takes 2·50 seconds and an unknown gas takes 10 seconds to penetrate through a porous surface. What is the density of the unknown gas with respect to hydrogen. *(Guru Nanak Pre-Univ., 1980)* **[Ans. 16]**

18. In 50 seconds, 300 ml of oxygen diffuse through a porous pot. How long will 500 ml of chlorine take to diffuse through the same pot under similar conditions ? (The molecular weights of chlorine and oxygen being known i.e., 71 and 32 respectively) **[Ans. 75 seconds]**

19. 5 litres of hydrogen (density=1) diffuse in 8 seconds. How long will it take 4 litres of oxygen (density=16) to diffuse under similar conditions ? *(Punjab H.S. 1971)* **[Ans. 25·6 second]**

20. 50 cc of hydrogen take 10 minutes to diffuse out of a vessel. How long will 40 cc of oxygen take to diffuse under similar conditions ? *(Himachal Pre-Univ., 1978)* **[Ans. 32 minutes]**

Diffusion of Gases

21. The reaction between gaseous NH_3 and gaseous HBr produces white solid NH_4Br. Suppose a small quantity of gaseous NH_3 and gaseous HBr were introduced simultaneously into the opposite ends of an open tube which is one metre long. Calculate the distance of the white solid formed from the end which was used to introduce ammonia.

[N = 14; Br = 80; H = 1] *(Kurukshetra Pre-Univ., 1975)*
[**Ans.** 68·6 cms]

22. In 100 seconds 300 ml of nitrogen diffuse through a porous partition. How long will 400 ml of CO_2 take to diffuse through the same partition under similar conditions?

(Guru Nanak Dev Pre-Univ., 1981)
[**Ans.** 167·14 seconds]

Volumes of Diffusion

23. If 100 volumes of air take 9 seconds to diffuse from a vessel, how long would it take the same volume of chlorine to do so under similar conditions? Densities of air and chlorine being 14·4 and 35·5 respectively?
[**Ans.** 14·10 seconds]

24. 8 ml of hydrogen were found to diffuse in 15 seconds. What volume of SO_2 (mol wt = 64) would diffuse in the same time under similar conditions? *(Punjab Pre-Univ., 1973)*
[**Ans.** 1·414 ml]

25. The relative rates of diffusion of oxygen and another gas A are in the ratio of 141·4 : 100. What is the molecular weight of A if that of oxygen is 32? *(Punjab Pre-Univ., 1976)*
[**Ans.** 64·00]

26. One litre each of nitrogen and hydrogen weigh 1·25 g and 0.09 respectively at NTP. Calculate the volume of nitrogen which would diffuse through a porous membrane in the same time as 100 ml of hydrogen at NTP. *(Himachal Pre-Univ., 1975)*
[**Ans.** 26·832 ml]

27. 15 ml of ozone diffuse through the porous pot in the same time as 12·50 ml of another gas A. Calculate the molecular weight of A. *(Himachal Pre-Univ., 1976)*
[**Ans.** 69·12]

28. Under similar conditions a gas diffuses twice as rapidly as CO_2. What is the molecular weight of the gas? *(All India Hr. Sec., 1979)*
[**Ans.** 11]

29. If 16·0 ml of hydrogen diffuse in 10 seconds, what volume of SO_2 will diffuse in the same time and under similar conditions? Molecular weight of SO_2 is 64. *(Punjab Board Hr. Sec., 1979 ml)*
[**Ans.** 4·0]

30. The relatively densities of sulphur dioxide and oxygen are 32 and 16 respectively. Calculate the volume of oxygen that will diffuse in the same time as 60 ml of sulphur dioxide.

(Aligarh Pre-Univ., 1980)
[**Ans.** 84·86]

31. Relative densities of carbon dioxide and oxygen are 21 and 16 respectively. If 25 cc of carbon dioxide diffuse in 75 seconds, what volume of oxygen will diffuse in 96 seconds under similar conditions? *(Punjab Pre Univ., 1982)*
[**Ans.** 37·44 cc]

8

Molecular Weights

MOLECULAR WEIGHTS OF GASSES

Avogadro's Hypothesis is helpful in finding the relative weight of the molecule (Molecular weight) of gaseous substances. According to **Avogadro's Hypothesis**, *"Equal volumes of all gases, under similar conditions of temperature and pressure, contain equal number of molecules"*. This means that—

(i) if we have 22·4 litres of different gases at S.T.P., they will have equal number of molecules in them.

(ii) the weight of 22·4 litres of any gas at S.T.P. will be equal to the sum total of the weights of all the molecules contained in 22·4 litres of this gas.

We know from the weight-volume relationship of gases that 22·4 litres of any gas at S.T.P. weighs equal to its molecular weight (in grams) called Gram-Molecular Weight **(G.M.W.)**. Now different gases have different molecular weights and since they all have same number of molecules in a definite volume under similar conditions of temperature and pressure, they must be differing in respect of the weights of their individual molecules. This means that the molecules of one gas must be lighter or heavier than that of the other. We shall now proceed to determine the **relative weight of one molecule of each gas**. This can be done by a number of methods, the important ones being.

Molecular Weights

(i) *The Molar-volume Method* ;

(ii) *The Vapour-density Method* ;

(iii) *By the application of Graham's Law of diffusion.*

I. Molar-volume method. The volume of one mole of any gas or gram-molecular weight of any gas at S.T.P. is called Molar-Volume or Gram-Molar-Volume (G.M.V.). The gram molecular weight of any gas occupies a definite and constant volume at standard conditions (S.T.P.). This fact forms an important principle in making chemical calculations regarding the molecular weight of a gas.

In this method, a sample of a gas is weighed and its volume measured at a particular temperature and pressure. This volume is carried to S.T.P by applying the gas equation. Mathematically, the weight of 22·4 litres of the gas is calculated which is the **approximate molecular weight of the gas.**

Example 1. *A sample of a gas weighing 0·775 g occupied 280 ml at S.T.P. Determine its approximate molecular weight.*

We are given that :

the weight of 280 ml of the gas at S.T.P. $= 0.775$ g

the weight of 1 ml of the gas at S.T.P. $= \dfrac{0.775}{280}$ g

and the weight of 22400 ml of the gas at S.T.P.

$$= \frac{0.775}{280} \times 22400 = 62 \text{ g/mol.}$$

Example 2. *A sample of a gas weighing 0·5366 g occupied a volume of 450 ml at 18°C and 720mm pressure. What is the approximate molecular weight ?*

STEP I

Find the volume of the given mass of gas at S.T.P.

Now,

$V_1 = 450$ ml

$P_1 = 720$ mm

$T_1 = 18 + 273$

$ = 291°$Abs.

$$P_2 = 760 \text{ mm}$$
$$T_2 = 273°\text{Abs.}$$
$$V_2 = ?$$

∴ Applying the gas equation

$$\frac{P_1V_1}{T_1} = \frac{P_2V_2}{T_2}, \text{ we have}$$

$$\frac{720 \times 450}{291} = \frac{760 \times V_2}{273}$$

or
$$V_2 = \frac{720 \times 450 \times 273}{291 \times 760} = 400 \text{ ml}$$

STEP II

Find the approximate Molecular Weight.

Now, 400 ml of the gas at S.T.P. weighs = 0·5366 g

$$1 \text{ ml of the gas at S.T.P. weighs} = \frac{0·5366}{400}$$

22400 ml of the gas at S.T.P. weighs

$$= \frac{0·5366}{400} \times 22400$$

$$= 30·04 \text{ g/mole}$$

II. Vapour density method. This method is applicable for volatile substances. *Vapour density of a substance is the "ratio between the weight of a certain volume of the vapour of the substance to the weight of the same volume of hydrogen gas, under similar conditions of temperature and pressure."*

Mathematically,

$$\text{V.D.} = \frac{\text{wt. of a certain vol. of the vapours}}{\text{wt. of the same vol. of hydrogen}} \text{ at S.T.P.}$$

We know that :

Mol. wt. $= 2 \times$ V.D.

Thus, if the vapour density of a substance is known, the molecular weight can be calculated.

Molecular Weights

Example 3. *0.8015 litre of chlorine at 15°C and 750 mm. pressure weigh 2.374 g. Find the molecular weight of chlorine.*

STEP I

Find the volume of chlorine at S.T.P.

Applying the gas equation,

$$\frac{P_1 V_1}{T_1} = \frac{P_2 V_2}{T_2}$$

and substituting the values we have,

$$\frac{750 \times 0.8015}{288} = \frac{760 \times V_2}{273}$$

$$V_2 = \frac{750 \times 0.8015 \times 273}{760 \times 288} = 0.75 \text{ litre}$$

STEP II

Find the vapour density.

$$\text{V.D.} = \frac{\text{wt. of 0.75 litre of chlorine}}{\text{wt. of 0.75 litre of } H_2} \text{ at S.T.P.}$$

$$= \frac{2.374}{0.75 \times 0.09} \text{ (wt. of 1 litre of } H_2$$

at S.T.P. $= 0.09$ gm.)

$$= 35.2$$

STEP III

Find the Mol. wt. of chlorine.

We know that:

$$\text{Mol. wt.} = 2 \times \text{V.D.}$$

∴ Mol. wt. of chlorine

$$= 2 \times 35.2$$

$$= 70.4$$

Example 4. *0.22 g of a substance when vaporised displaced 45.0 ml of air measured at 29°C and 755 mm pressure.*

Calculate the molecular weight of the substance. (Aq. tension at 29°C is 17·5 mm).

STEP I

Find the volume of air displaced at S.T.P.

P_1 = 755 mm — Aq. tension 17·5 mm
   = 737·5 mm
V_1 = 45·0 ml
T_1 = 302° Abs.
P_2 = 760 mm
V_2 = ?
T_2 = 273° Abs

Substituting the values in the gas equation

$$\frac{P_1 V_1}{T_1} = \frac{P_2 V_2}{T_2}$$

$$\frac{737 \cdot 5 \times 45}{302} = \frac{760 \times V_2}{273}$$

or $\quad V_2 = \dfrac{737 \cdot 5 \times 45 \times 273}{760 \times 302} = 39 \cdot 47$ ml

STEP II

Find the vapour density.

$$V.D = \frac{\text{wt. of 39·47 ml of the air}}{\text{wt. of 39·47 ml of } H_2} \text{ at N.T.P.}$$

$$= \frac{0 \cdot 22}{39 \cdot 47 \times 0 \cdot 00009} = 61 \cdot 93$$

(wt of 1 ml of H_2 at N.T.P. = 0·00009 g)

STEP III

Find the mol. wt.

Mol. wt. = 2 × V.D.
    = 2 × 61·93 = 123·86.

Molecular Weights

III Application of Graham's Law of diffusion. This method has been dealt with in detail in the previous chapter.

Density of gases from molecular weights. Density is mass per unit volume. If we know the mass of 1 ml of the gas we can find its density by dividing its mass by volume,

i.e., $$\text{Density} = \frac{\text{Mass}}{\text{Volume}}$$

One ml is a very small volume. For the sake of convenience 1 litre has been taken as the unit of volume. Thus when we say that the density of hydrogen is 0.0889 at S.T.P., we mean that the mass of 1 litre of hydrogen at S.T.P. is 0.0889 g. Similar is the case with other substances.

Knowing this that density of a gas is the mass of 1 litre of the gas at S.T.P., we can easily calculate the density of the gas if its molecular (Mole-weight) weight is known. The molecular weight of a substance can be found by many methods, the simplest being by adding the atomic weight of all the component atoms when the formula is known.

From the weight-volume relationship we know that 22.4 litres of every gas weigh equal to its mol. wt. in grams at S.T.P. Thus we can find the density of the substance by dividing the molecular weight of the substance by 22.4.

Example 5. *The molecular weight of nitric oxide is 30 0. Calculate its density.*

Molecular weight of nitric oxide = 30.00

This means that 22.4 litres of nitric oxide at S.T.P. weigh

$$= 30.0 \text{ g (G.M.W.)}$$

∴ 1 litre of nitric oxide at S.T.P. will weigh

$$= \frac{30}{22.4} = 1.34 \text{ g}$$

Hence density of nitric oxide = 1.34 g/litre

Example 6. *Calculate the density of CO_2 at S.T.P.*

Step I

Find the Mol. wt. of CO_2.

$$\text{Wt. of carbon} = 1 \times C = 1 \times 12 = 12$$
$$\text{Wt. of oxygen} = 2 \times O = 2 \times 16 = 32$$
$$\therefore \quad \text{Mol. wt. of } CO_2 = 44$$

Step II

Find the weight of one litre of CO_2.

Now 22.4 litres of CO_2 at S.T.P. weigh = 44 g

$$1 \text{ litre of } CO_2 \text{ at S.T.P. weighs} = \frac{44}{22 \cdot 4} = 1 \cdot 95 \text{ g}$$

Hence, the density of $CO_2 = 1 \cdot 95$ g/litre.

Example 7. *Find the density of a gaseous substance if 280 ml weigh 0·55 g (at S.T.P.).*

The weight of 280 ml at S.T.P. = 0·55 g

The weight of 1 ml at S.T.P. = $\dfrac{0 \cdot 55}{280}$ g

$\therefore$ the weight of 1 litre at S.T.P.

$$= \frac{0 \cdot 55}{280} \times 1000 = 1 \cdot 964 \text{ g}$$

Thus, the density = 1·964 g/litre.

END-OF-CHAPTER PROBLEMS

Molar-Volume Method

1. A volatile substance weighing 0·125 g displaced 31·0 ml of dry air in a Victor Meyer's determination at 100°C and 735 mm pressure. Calculate the molecular weight of the substance. [Ans. 175]

2. Find the weight of two litres of ammonia at S.T.P. [Ans. 1·520 g]

3. What weight of hydrogen at S.T.P. could be contained in a vessel which holds 3·0 g of oxygen at S.T.P. ? [Ans. 0·1875 g]

4. 0·135 g of a substance when vaporised displaced 28·8 ml of air at 15°C and 762·7 mm pressure. Calculate the molecular weight of the substance. (Aq. tension at 15°C is 12·7 mm). [Ans. 112·34]

5. The weight of 2 litres of oxygen at S.T.P. is 2·85 g. What is the weight of 0·5 litre of oxygen at 290°Abs. and 770 mm. pressure ?
[Ans. 0·7055 g]

6. The weight of equal volumes of CO_2 and HCl measured under similar conditions were 0·547 g and 0·660 g respectively. The molecular weight of HCl is 36·5, find that of CO_2. [Ans. 44]

7. 0·056 g of a substance, on vaporisation in a Victor Meyer's apparatus displaced 19·0 ml of moist air at 27°C and 746·5 mm pressure. Calculate the molecular weight of the substane. (Aqueous tension at 27°C = 26·7 mm). [Ans. 76·56]

8. What weight of barium peroxide is required for the liberation of 50 litres of oxygen at 15°C and 710 mm pressure ? [Ans. 670 g]

9. 0·823 g of a hydrocarbon occupied 617 ml at 684 mm pressure and 27°C. It contains 79·88% carbon and 20·12% hydrogen. What is the molecular weight and molecular formula of the substance ?
[Ans. 30 ; C_2H_6]

10. What is the molecular weight of a gas, 1·12 litres of which at 0°C and 760 mm pressure weigh equal to 3·2 gms ? What is the vapour density of the gas ? *(Punjabi Pre-Univ., 1973)*
[Ans. 64, 32]

11. Calculate the volume at N.T.P. occupied by

(i) 16 g of oxygen.

(ii) 1·5 moles of nitrogen.

(iii) $6·023 \times 10^{25}$ molecules of CO_2.

(Guru Nanak Dev Pre-Univ, 1975)
[Ans. (i) 11200 ml (ii) 33600 ml (iii) 22400 ml]

12. Calculate the volume at N.T.P. occupied by :

(i) 0·5 mole of nitrogen.

(ii) $6·023 \times 10^{23}$ molecules of hydrogen sulphide.

(iii) one gram equivalent of oxygen.

(Punjabi Pre-Univ., 1980)
[Ans. (i) 11·2 litres ; (ii) 22·4 litres ; (iii) 5·6 litres]

13. What is the number of molecules in 22·4 litres of nitrogen at N.T.P. ? *(Punjab Pre-Univ., 1980)*
[Ans. $6·023 \times 10^{23}$ molecules]

14. Correct and rewrite the following statement : "22·4 litres of O_2 have weight 32 g". *(Guru Nanak Dev Pre-Univ 1980)*
[Ans. 22·4 litres of O_2 at N·T.P. have weight 32 g]

15. What is the volume occupied by 2·50 moles of CO_2 at N.T.P. ?
(Punjabi Pre-Univ., 1980)
[Ans. 56 litres]

16. How much volume will be occupied by 2 g of dry oxygen at 27°C and 740 mm pressure ? *(Guru Nanak Dev Pre-Univ., 1981)*
[**Ans.** 1·58 litres]

17. What will be the volume of gas at 17°C and 730 mm pressure if it occupies a volume of 22·4 litres at N.T.P. *(Punjabi Pre-Univ., 1981)*
[**Ans.** 6763 litres]

18. What will be the weight of 150 cc of oxygen collected over water at 15°C and 740 mm pressure ? (Aq. Tension at 15°C = 13 mm). *(Punjabi Pre-Univ., 1982)*
[**Ans.** 0·194 gm]

19. 0·150 gm. of a volatile substance when treated in Victor Meyer's apparatus displaced 40·50 cc of air collected over water at 15°C and 746 mm pressure. Calculate the molecular weight of the substance. *(Maharshi Dayanand Pre-Univ., 1982)*
[**Ans.** 90·83]

Vapour Density

20. The vapour density of a gaseous substance is 40·1. Find the molecular weight. [**Ans.** 80·2]

21. In a Victor-Meyer's experiment, the following data was noted by a student :

Wt. of the sample taken = 1·008 g
Vol. of air displaced = 220·00 ml
Temperature = 16·5°C
Barometer reading = 707·5 mm
Aq. tension at 16·5°C = 14·0 mm

Determine the vapour density and molecular weight of the sample.
[**Ans.** 59·5 ; 119]

22. In a Victor-Meyer's apparatus, 0·243 g of a volatile substance displaced 35 ml of air at 23°C and 745 mm pressure. Calculate the molecular weight and vapour density of the substance.
[**Ans.** V.D. = 88·95, Mol. wt. = 177·9]

23. 0·6 g of a volatile substance displaced 123·0 ml of moist air at 20°C and 757·4 mm pressure. Calculate the molecular weight of the volatile substance. (Aq. tension at 20°C = 17·4 mm).
[**Ans.** V.D. = 60·2, Mol. wt. = 120·4]

24. In a vapour density determination by Victor-Meyers's method, 0·1348 g of a volatile liquid displaced 25·8 ml of air measured at N.T.P. Calculate the V.D. and mol. wt. of the substance.
[**Ans.** V.D. = 58·05 ; Mol. wt. = 116·1]

25. 0·168 gm of a substance when vaporised displaced 49·4 c.c. of air at 20°C and 740 mm pressure. Calculate its molecular weight. (Aq. tension at 20°C is 1·8 cm and weight of 1 c.c. of H_2 at N.T.P. is 0·00009 g). *(U.P. Board Inter., 1974)*
[**Ans.** 84·6]

26. In Victor-Meyer's determination 0·3 gm of a substance displaced 60 c.c. of air at 27°C and 757 mm pressure. Calculate the molecular weight of the substance. (Aq. tension at 27°C = 27mm ; weight of 1 c.c. of hydrogen at N.T.P. = 0·00009 gm). *(Rajasthan Pre-Univ., 1974)*
[**Ans.** 127]

27. 0·2 gm of a volatile substance displaced 30 ml of air at 27°C. and 760 mm pressure in Victor-Meyer's method. Calculate the vapour

Molecular Weights 91

density and molecular weight of the substance (1 ml of H_2 at N.T.P. weighs 0·00009 gm). *(Guru Nanak Dev Pre-Univ., 1975)*
[Ans. 81·4, 162·8]

Density

28. Calculate the density of methane (CH_4) at S.T.P.
[Ans. 0·714 g/litre]

29. Calculate the density of sulphur dioxide, H_2S and NH_3 at S.T.P. [Ans. 2·85 g/litre, 1·51 g/litre and 0·75 g/litre]

30. 400 ml of a gas at S.T.P. weigh 0·540 g. Find the density at S.T.P. [Ans. 1·35 g/litre]

31. 0·6 g of a volatile substance displaced 123 cc of moist air at 20°C and 157·4 mm pressure. Calculate the V.D. of the gas. (Aqueous tension at 20°C is 17·5 mm). [Ans. 120·5]

32. 0·27 g of a volatile liquid displaced 750 ml of dry air at N.T.P. Calculate the vapour density of the liquid. [Ans. 4·1]

33. A storage tank contains 20 litres of dry oxygen at 27°C and 60 atmosphere pressure. Calculate the weight of oxygen gas in tank.
[Ans. 1560 g]

34. 0·406 g of a volatile substance in Victor-Meyer's apparatus expelled 170 cc of air measured over water at 15°C and 763 mm pressure. Find out the molecular weight of the substance. (Aqueous tension at 15°C = 13 mm) *(Marathwada Pre-Univ., 1975)*
[Ans. 57·11]

35. In a Victor-Meyer's experiment 0·23 g of a substance displaced air which measured 112 ml at N.T.P. Calculate the molecular weight of the substance. *(Gujarat Pre-Univ., 1975)*
[Ans. 46·00]

36. At 17°C and 735 mm pressure 615 mm of a gas weighed 2·75 g. What is the molecular weight of the gas ? *(Andhra Pre-Univ., 1976)*
[Ans. 110·01]

37. Suppose that you have discovered a new gaseous compound and find that at N.T.P. 100 ml of it weigh 0·32 g. What will be its molecular weight ? *(All India Hr. Sec., 1976)*
[Ans. 71·68]

38. 0·1 g of a volatile substance on vaporisation in Victor-Meyer's apparatus displaced 27 ml of moist air measured at 15°C and 740 mm pressure. Calculate the molecular weight of the subtance. (Aqueous tension at 15°C = 12·7 mm).
(Guru Nanak Dev Pre-Univ., 1977)
[Ans. 92·4]

39. In a Victor-Meyer's experiment 0·6 g of a volatile substance displaced 123 ml of moist air at 20°C and 757·4 mm pressure. Calculate the vapour density and molecular weight of the substance. (Aqueous tension at 20°C = 17·4 mm and wt. of 1 ml of hydrogen at N.T.P. = 0·00009 g.) *(Punjab Board Hr. Sec., 1977)*
[Ans. V.D. = 60·25 ; Mol. wt. = 120·5]

9

Equivalent Weights

GRAM-EQUIVALENT WEIGHT

The equivalent weight of an element has been defined as that amount of it which will combine with or displace from 1·008 parts by weight of hydrogen, 8 parts by weight of oxygen and 35·5 parts by weight of chlorine.

This definition immediately gives us a *reference standard* for measuring the combining capacity of elements with respect to each other. Since we have to adopt some unit of mass for this reference standard, the term **Gram-equivalent weight** has been used. Gram-equivalent weight of an element may be defined as *that quantity of the element in grams which reacts with or displaces 1 gram-atom i.e., 1·008 grams of hydrogen or 1 gram-atom of oxygen i.e., 8 grams of oxygen or 1 gram atom of chlorine i.e, 35·5 grams of chlorine.*

Equivalent Weight and Valency. If one gram atomic weight of an element reacts with 1 gram-atom or 1·008 grams of hydrogen, we say that the combining capacity or valency of the element is the *same* as that of hydrogen, *i.e.*, one. Similarly, if one gram-atom of an element combines with 2 gram-atoms of hydrogen ($2 \times 1·008$ g), the valency of this element is twice that of hydrogen, *i.e.*, two. Thus we find that gram-atomic weight, gram-equivalent weight and valency of an element are related as :

$$\frac{\text{Gram atomic weight}}{\text{Gram equivalent weight}} = \text{Valency}$$

METHODS FOR FINDING EQUIVALENT WEIGHTS

A number of methods are employed for the determination of equivalent weights of elements. These are :

1. Hydrogen Displacement method. This method is suitable for metals which lie above hydrogen in the electrochemical series *e.g.*, Ca, Zn, Fe, Cd, Co, Sn, Pb, etc., which liberate hydrogen gas when treated with dilute acids like HCl or H_2SO_4. From the weight of the metal taken and the volume of hydrogen gas evolved experimentally, the equivalent weight of the metal can be determined.

Calculation. Let the weight of the metal taken = w g.

Let the vol. of hydrogen collected = V_1 ml

at $t_1°C$ and p_1 mm atmospheric pressure.

Let the aq. tension at $t_1°C = p'$ mm.

∴ pressure of the dry gas = $p_1 - p' = p$ mm.

Applying the gas equation we can find the volume of hydrogen at S.T.P. Let it be V ml.

Since the weight of 1 ml of H_2 at S.T.P. = 0·00009 g.

∴ wt. of V ml of hydrogen at S.T.P. = $V \times 0·00009$ g.

Now $V \times 0·00009$ g of the hydrogen is evolved by metal

$$= w \text{ g}$$

∴ 1 ,, ,, will be evolved by metal

$$= \frac{w}{V \times 0·00009}$$

Thus the gram-equivalent weight of the metal concerned

$$= \frac{\text{wt. of the metal}}{\text{vol. of } H_2 \text{ at S.T.P.} \times 0·00009}$$

Alternatively since 22400 ml of H_2 at S.T.P. weigh = 2 g

∴ 11200 ml ,, ,, will ,, = 1 g

Thus v ml of H_2 (at S.T.P.) are evolved from metal

$$= w \text{ g}.$$

∴ 11200 ml of H_2 (S.T.P.) will be evolved from metal

$$= \frac{w}{v} \times 11200 \text{ g}$$

This is the Gram-equivalent weight of the metal.

Example 1. *In an experiment 0.5 g of metal displaced exactly 0.0218 g of hydrogen. What is the equivalent weight of the metal ?*

Here we are given the weight of hydrogen gas evolved.

Now 0.0218 g of hydrogen is obtained from metal

$$= 0.5 \text{ g}$$

∴ 1 g of hydrogen will be obtained from metal

$$= \frac{0.5}{0.0218} = 22.93$$

Thus the Gram-equivalent of the metal = 22.93 g.

Example 2. *A quantity of zinc weighing 0.127 g when treated with excess hydrochloric acid, yielded 48.2 ml of hydrogen at 21°C and 758 mm pressure. Find the equivalent weight and gram-equivalent weight of zinc. Aqueous tension at 21°C is 18.5 mm.*

We are given :

Weight of zinc = 0.127 g

Vol. of H_2 evolved = 48.2 ml

Temperature = 21°C

Pressure of moist gas = 758 mm

Pressure of vapours of water (Aq. tension) = 18.5 mm

STEP I

Find the volume of H_2 at S.T.P.

$P_1 = 758 - 18.5 = 739.5$
$V_1 = 48.2$ ml
$T_1 = 273 + 21 = 294°$ Abs.
$P_2 = 760$ mm

Equivalent Weights

$$V_2 = ?$$
$$T_2 = 273° \text{ Abs.}$$

Applying gas equation

$$\frac{P_1V_1}{T_1} = \frac{P_2V_2}{T_2}, \text{ we have}$$

$$\frac{739 \cdot 5 \times 48 \cdot 2}{294} = \frac{760 \times V_2}{273}$$

$$\therefore \quad V_2 = \frac{48 \cdot 2 \times 739 \cdot 5 \times 273}{760 \times 293} = 43 \cdot 7 \text{ ml.}$$

Step II

Find the weight of 43·7 ml of H_2

Now the weight of 1 ml of H_2 at S.T.P. = 0·00009 g

∴ the weight of 43·7 ml of H_2 at S.T.P.
$$= 0 \cdot 00009 \times 43 \cdot 7 = 0 \cdot 0039 \text{ g}$$

Step III

Find the equivalent weight.

Now 0·0039 g of H_2 is obtained from zinc = 0·127 g

∴ 1 g of H_2 is obtained from zinc

$$= \frac{0 \cdot 127}{0 \cdot 0039} = 32 \cdot 6 \text{ g}$$

Thus the gram-equivalent weight of zinc = 32·6 g

and the Equivalent weight = 32·6

2. Oxide formation method For such metals which readily combine with oxygen to form stable oxides this method is most suitable. The metals like iron, zinc, magnesium, mercury all form stable oxides. In this method, a known weight of the metal is heated in air or oxygen till constant weight. The weight of metal oxide thus formed is recorded.

Calculations. Let the weight of metal taken = w_1 g

Let the wt. of oxide formed = w_2 g

∴ wt. of oxygen combining with w_1 g metal $= w_2 - w_1$ g

Now $w_2 - w_1$ g oxygen combines with metal $= w_1$ g

∴ 1 g oxygen combines with metal $= \dfrac{w_1}{w_2 - w_1}$ g

or 8 g oxygen combines with metal $= \dfrac{w_1}{w_2 - w_1} \times 8$ g

Hence eq. weight of the metal $= \dfrac{\text{wt. of the metal}}{\text{wt. of oxygen}} \times 8$.

Example 3. *When 1·1338 g of aluminium is heated in an atmosphere of oxygen, the resulting aluminium oxide is 2·1428 g. Calculate the equivalent weight and valency of aluminium.*

We are given that :

$$\text{Weight of Al} = 1\cdot1338 \text{ g}$$

Weight of aluminium oxide formed $= 2\cdot1428$ g

∴ wt. of O_2 combining with 1·1338 g Al

$$= 2\cdot1428 - 1\cdot1338 = 1\cdot009$$

Now 1·009 g oxygen combines with Al $= 1\cdot1338$

∴ 8 g oxygen will combine with Al

$$= \dfrac{1\cdot1338}{1\cdot009} \times 8 = 8\cdot99$$

We know that valency $= \dfrac{\text{Atomic weight}}{\text{Equivalent weight}}$

Hence, valency $= \dfrac{27}{8\cdot99} = 3$.

Example 4. *Aluminium oxide contains 47·19% oxygen. Determine the equivalent weight of aluminium*

We are given that :

100 g of aluminium oxide contains oxygen $= 47\cdot19$ g

∴ $\qquad$ Wt. of Al $= 100 - 47\cdot19 = 52\cdot81$ g

Equivalent Weights

Now 47·19 g of oxygen combines with Al = 52·81 g

∴ 1 g of oxygen combines with Al $= \dfrac{52 \cdot 81}{47 \cdot 19}$

So, 8 g of oxygen combines with Al $= \dfrac{52 \cdot 81}{47 \cdot 19} \times 8$

$$= \dfrac{422 \cdot 48}{47 \cdot 19} = 8 \cdot 95$$

Example 5. *A sample of copper weighing 0·8013 g was dissolved in nitric acid to form copper nitrate. This copper nitrate was then heated until it was converted to copper oxide. The weight of copper oxide was found to be 1·0035 g. What is the equivalent weight of copper ?*

In this case, oxide of copper is obtained *indirectly*. The metal is first converted to its nitrate which on further heating gives its oxides.

Copper + nitric acid ⟶ Copper nitrate

Copper nitrate ⟶ Copper oxide

We are given :

Weight of metal oxide = 1·0035 g

Weight of metal taken = 0·8013 g

∴ Wt. of oxygen combining with 0·8013 g metal

$$= 1 \cdot 0035 - 0 \cdot 8013 = 0 \cdot 2022 \text{ g}$$

Now 0·2022 g of oxygen combines with metal = 0·8013 g

1 g of oxygen combines with metal $= \dfrac{0 \cdot 8013}{0 \cdot 2022}$ g

8 g of oxygen combines with metal $= \dfrac{0 \cdot 8013}{0 \cdot 2022} \times 8$ g

$$= \dfrac{6 \cdot 4104}{0 \cdot 2022} = 31 \cdot 7 \text{ g}$$

Hence the equivalent weight of copper = 31·7.

3. Reduction of oxide method. For such metals whose oxide can be easily reduced, this method has been found to be quite suitable. The oxides of the metals like zinc, iron, lead, mercury can be reduced to give corresponding metals.

In this method, a known weight of the metal oxide is heated in the atmosphere of hydrogen till constant weight. From the weight of the metal oxide taken and that of the metal left after reduction, equivalent weight can be found.

Calculations. Let the weight of the metal oxide $= w_1$ g

Let the wt. of the metal left $= w_2$ g

∴ Wt. of oxygen combining with w_2 g metal
$$= (w_1 - w_2) \text{ g}$$

Now $(w_1 - w_2)$ g oxygen combines with metal $= w_2$ g

∴ 1 g of oxygen combines with metal $= \dfrac{w_2}{w_1 - w_2}$ g

and 8 g of oxygen combines with metal $= \dfrac{w_2}{w_1 - w_2} \times 8$ g

∴ eq. wt. of the metal $= \dfrac{\text{wt. of the metal}}{\text{wt. of oxygen}} \times 8$.

Example 6. *A sample of mercury oxide weighing 1·7383 g was heated in an atmosphere of hydrogen, and 1·6716 g of mercury metal was left after the reaction. Find the equivalent weight of mercury.*

We are given :

Wt. of mercury oxide $= 1·7383$ g

Wt. of mercury metal $= 1·6716$ g

∴ Wt. of oxygen combining with 1·6716 g of mercury
$$= 1·7383 - 1·6716 = 0·0667 \text{ g}$$

Now 0·0667 g oxygen combines with mercury
$$= 1·6716 \text{ g}$$

∴ 1 g oxygen combines with mercury $= \dfrac{1·6716}{0·0667}$ g

and 8 g oxygen combines with mercury $= \dfrac{1·6716}{0·0667} \times 8$ g

$$= 200·4 \text{ g}$$

Hence the eq. wt. of mercury $= 200·4$

Equivalent Weights 99

Example 7. *On analysis it was found that 8·10 g of an oxide of iron contains 5·66 g iron. What is the equivalent weight of iron?*

We are given :

Weight of iron oxide = 8·10 g

Weight of iron metal = 5·66 g

∴ Wt. of oxygen combining with 5·66 g iron = 2·44 g

Now 2·44 g of oxygen combines with iron = 5·66 g

∴ 1 g of oxygen combines with iron = $\dfrac{5·66}{2·44}$ g

8 g of oxygen combines with iron = $\dfrac{5·66}{2·44} \times 8$

$$= 18·55 \text{ g}$$

Hence eq. wt. of iron = 18·55.

4. Chloride formation method. This method is especially suitable for metals like gold and silver which readily combine with chlorine to form their corresponding chlorides.

$$\text{Metal} + \text{chlorine} \longrightarrow \text{Metal chloride}$$

A known weight of the metal is treated with chlorine gas till constant weight. From the weight of the metal taken and that of the metal choride formed, eq. wt. of the metal can be calculated.

Calculations. Let the weight of the metal = w_1 g

Let the wt. of the metal chloride = w_2 g

∴ Wt. of chlorine combining with w_1 g metal

$$= (w_2 - w_1) \text{ g}$$

Now $(w_2 - w_1)$ g chlorine combines with metal = w_1 g

1 g chlorine combines with metal = $\dfrac{w_1}{w_2 - w_1}$ g

∴ 35·5 g chlorine combines with metal

$$= \dfrac{w_1}{w_2 - w_1} \times 35·5 \text{ g}$$

Hence eq. wt. of the metal = $\dfrac{\text{Wt. of the metal}}{\text{Wt. of the chlorine}} \times 35·5$

Example 8. *In an experiment it was found that 4·8617 g of chlorine combined with an element to give 4·9999 g of a chloride. What is the equivalent weight of the element?*

We are given :

Weight of chloride = 4·9999 g

Weight of chlorine = 4·8617 g

Wt. of the element in the chloride = 0·1382 g

Now 4·8617 of chlorine combines with element

$$= 0 \cdot 1382 \text{ g}$$

1 g of chlorine combines with element

$$= \frac{0 \cdot 1382}{4 \cdot 8617} \text{ g}$$

35·5 g of chlorine combines with element

$$= \frac{0 \cdot 1382}{4 \cdot 8617} \times 35 \cdot 5 = 1 \cdot 01 \text{ g}$$

Hence eq. wt. of the element = 1·01.

Example 9. *The analysis of a carefully prepared chloride of an element showed it to have 24·74% chlorine, the rest being the element itself. Find the equivalent weight of the element.*

We are given :

Wt. of chloride = 100 g

Wt. of chlorine = 24·74 g

∴ Wt. of element = 75·26 g

Now 24·74 g of chlorine combines with element

$$= 75 \cdot 26 \text{ g}$$

1 g of chlorine combines with element

$$= \frac{75.26}{24 \cdot 74} \text{ g}$$

∴ 35·5 g of chlorine will combine with element

$$= \frac{75 \cdot 26}{24 \cdot 74} \times 35 \cdot 5 = 107 \cdot 99 \text{ g}$$

Hence the eq. wt. of the element = 107·99

5. Double Decomposition Method.

This method involves the mixing up of the solutions of the two salts. The weight of one of the salts is known. Due to ionisation, the salts split up into ions and exchange partners in solution involving what is commonly known as *double decomposition*. The precipitate of one of the products of the reaction is usually formed. The precipitate is filtered, washed, dried and weighed.

$$AB + CD \longrightarrow AD + CB.$$

Calculations. Let the weight of AB $= w_1$ g

Let the wt. of AD (ppt.) $= w_2$ g

According to the law of equivalents,

$$\frac{\text{Wt. of AB}}{\text{Wt. of AD}} = \frac{\text{eq. wt. of AB}}{\text{eq. wt. of AD}}$$

Thus knowing the equivalent weight of either of the two (AB or AD), the equivalent weight of the other can be found out.

This method also enables us to find the equivalent weight of the individual ions or radicals. Thus :

$$\frac{\text{Wt. of AB}}{\text{Wt. of AD}} = \frac{\text{eq. wt. of A} + \text{eq. wt. of B}}{\text{eq. wt. of A} + \text{eq. wt. of D}}$$

If the equivalent weight of any two of A, B and D are known, that of the third can be calculated.

Example 10. *0·194 g of a metallic chloride, on treatment with excess of silver nitrate, gave a precipitate of 0·5 g silver chloride, calculate the equivalent weight of the metal.*

Metal chloride $\xrightarrow[\text{AgNO}_3]{\text{Treatment with}}$ Metal nitrate + silver chloride

Weight of metal chloride $= 0·194$ g

Weight of silver chloride $= 0·5$ g

Applying the law of equivalents,

$$\frac{\text{Wt. of metal chloride}}{\text{Wt. of silver chloride}} = \frac{\text{Eq. wt. of metal} + \text{eq. wt. of chlorine}}{\text{Eq. wt. of silver} + \text{eq. wt. of chlorine}}$$

or $\dfrac{0\cdot 194}{0\cdot 5} = \dfrac{\text{Eq. wt. of the metal (say } E) + 35\cdot 5}{108 + 35\cdot 5}$

or $\dfrac{0\cdot 194}{0\cdot 5} = \dfrac{E + 35\cdot 5}{143\cdot 5}$

or $0\cdot 5\,(E + 35\cdot 5) = 143\cdot 5 \times 0\cdot 194$

or $0\cdot 5\,E + 17\cdot 75 = 143\cdot 5 \times 0\cdot 194$

or $0\cdot 5\,E = 143\cdot 5 \times 0\cdot 194 - 17\cdot 75$

$\therefore \quad E = \dfrac{143\cdot 5 \times 0\cdot 194 - 17\cdot 75}{0\cdot 5} = 20\cdot 178.$

6. Metal displacement method. It has been observed that a metal lying higher in electro-chemical series to the other can displace the latter from its salt solution. Thus zinc can displace copper from the solutions of copper salts or calcium can displace zinc from the solutions of the salts of zinc and so on. It is found that one metal displaces the other in the ratio of their equivalent weights. This means that 32·5 g of zinc (1 g equivalent weight of zinc) can displace 31·8 g (1 g equivalent weight of copper) or 108 g Ag can displace 32·5 g of zinc and so on. This fact has been found useful in finding the equivalent weight of one metal if that of the other is known.

Calculations. Let a metal A displace another metal B from the solution of the salt of B (say BX).

Let the wt. of A added $\quad = w_1$ g

Let the wt. of B displaced $\quad = w_2$ g

Let the eq. wt. of A $\quad = E$ (known)

Applying the law of equivalents

$$\dfrac{\text{Wt. of A}}{\text{Wt. of B}} = \dfrac{\text{Eq. wt. of A}}{\text{Eq. wt. of B}}$$

or $\quad \dfrac{w_1}{w_2} = \dfrac{E}{\text{Eq. wt. of B}}$

Thus Eq. wt. of $B = \dfrac{E \times w_2}{w_1}$

Equivalent Weights

Example 11. *0·181 g of magnesium displaced 1·612 g of silver from its salts. Find the equivalent weight of magnesium when that of Ag is 108.*

We are given :

Wt. of magnesium   = 0·181 g

Wt. of silver displaced   = 1·612 g

Applying the law of equivalents

$$\frac{\text{Wt. of magnesium}}{\text{Wt. of silver}} = \frac{\text{Eq. wt. of magnesium}}{\text{Eq. wt. of silver}}$$

or $\quad \dfrac{0\cdot 181}{1\cdot 612} = \dfrac{E}{108}$

or $\quad E = \dfrac{0\cdot 181 \times 108}{1\cdot 612} = 12\cdot 12$

Thus eq. wt. of magnesium = 12·12.

Example 12. *1·201 g of zinc gave 1·497 g of zinc oxide on treatment with nitric acid and subsequent ignition. In a second experiment 0·543 g of zinc precipitated 0·527 g of copper from a solution of copper sulphate. Calculate the equivalent weight of zinc assuming that of oxygen to be 8.*

We are given :

Wt. of zinc metal   = 1·201 g

Wt. of zinc oxide formed   = 1·497 g

STEP I

Find the eq. wt. of zinc from the above data.

Wt. of oxygen which combines with 1·201 g zinc
$$= 1\cdot 497 - 1\cdot 201 = 0\cdot 296 \text{ g}$$

Now 0·296 g of oxygen combines with zinc = 1·201 g

∴ 8 g of oxygen combines with zinc

$$= \frac{1\cdot 201}{0\cdot 296} \times 8 = 32\cdot 46 \text{ g}$$

Step II

Find the eq. wt. of copper.

Applying the Law of equivalents

$$\frac{\text{Wt. of zinc}}{\text{Wt. of copper}} = \frac{\text{Eq. wt. of zinc}}{\text{Eq. wt. of copper}}$$

$$\frac{0\cdot 543}{0\cdot 527} = \frac{32\cdot 46}{E}$$

or $\qquad E = \dfrac{32\cdot 46 \times 0\cdot 527}{0\cdot 543} = 31\cdot 5$

Thus, the eq. wt. of copper is 31·5.

7. Faraday's Electrolytic method. This method involves the application of both the laws of electrolysis. The equivalent weight of an element is known if its electro-chemical constant (or equivalent) is known. This method is based upon Faraday's first law of electrolysis. The equivalent weight of an element can also be found out with the help of of Faraday's second law of electrolysis.

(a) Faraday's First Law of Electrolysis. This law states that *"the amounts of substances decomposed as a result of electrolysis are directly proportional to the quantity of electricity passed."*

Expressed mathematically

$$w \propto Q$$

where w is the amount of the substance decomposed and Q is the quantity of electricity passed. Q is equal to the product of the current strength (c) and the time (t) for which current is passed

i.e., $\qquad Q = c \times t$

So that

$$w \propto c \times t$$

or $\qquad w = k \times c \times t$

where 'k' is the electro-chemical equivalent of the substance

when $c = 1$ ampere, $t = 1$ second,

$$w = k \times 1 \times 1$$
$$= k$$

Equivalent Weights

Thus 'k', *the electro-chemical equivalent of a substance is that amount of it which is decomposed by a current of 1 ampere strength in one second (or one coulomb electricity).*

It has been observed that when 96,500 coulombs of electricity are passed through the solution of an electrolyte, the quantity of the substance decomposed is always equal to its equivalent weight in grams. Thus when 96,500 coulombs of electricity is passed through acidulated water, the quantities of oxygen and hydrogen evolved at the electrodes will be 8 g and 1 g respectively (*i.e.*, their gram equivalent weights). It is thus clear that the equivalent weight of a substance is equal to its electro-chemical equivalent multiplied by 96,500

or
$$E = k \times 96{,}500$$

Thus, if we know the value of k, that of E can be found out.

Example 13. *A current of 5·0 amperes flowing for exactly 15 minutes deposited 1·524 g of zinc at the cathode. What is the equivalent weight of the metal?*

STEP I

Find the quantity of electricity passed.

$$c = 5 \cdot 0 \text{ amperes}$$
$$t = 15 \times 60 \text{ seconds}$$
$$\therefore Q \text{ (quantity of electricity)} = 5 \cdot 0 \times 15 \times 60$$
$$= 4500 \text{ coulombs.}$$

STEP II

Find the equivalent weight.

Now 4500 coulombs deposit zinc $= 1 \cdot 524$ g

$\therefore$ 1 coulomb will deposit zinc $= \dfrac{1 \cdot 524}{4500}$

and 96,500 coulombs will deposit zinc $= \dfrac{1 \cdot 524}{4500} \times 96{,}500$

Hence eq. wt of zinc $= 32 \cdot 7$.

8. Faraday's Second Law of Electrolysis. The law states that *"when same current is passed through different electrolytes for the same time, the amounts of the substances obtained as a result of electrolysis are in the ratio of their equivalent weights."*

Thus if we have two cells containing copper sulphate, zinc sulphate solution and same current is passed through these two cells connected in series for the same time, then according to the law

$$\frac{\text{Wt. of copper}}{\text{Wt. of zinc}} = \frac{\text{eq. wt. of copper}}{\text{eq. wt. of zinc}}$$

Example 14. *The same quantity of current that liberated 1·079 g of silver was passed through copper sulphate solution and deposited 0·318 g of copper. Find equivalent weight of copper if that of silver is 108.*

According to Faraday's Second Law

$$\frac{\text{Wt. of silver}}{\text{Wt. of copper}} = \frac{\text{eq. wt. of silver}}{\text{eq. wt. of copper}}$$

Substituting the values, we have

$$\frac{1 \cdot 079}{0 \cdot 318} = \frac{108}{E}$$

$$\therefore \quad E = \frac{10 \times 0 \cdot 318}{1 \cdot 079} = 31 \cdot 8$$

Thus, the equivalent weigh of copper is 31·8.

Example 15. *An electric current is passed through a solution of copper sulphate and silver cyanide connected in series. If in a given time 0·35 g of copper is deposited, what will be the weight of silver deposited in the same time ? At wt. of Cu = 63·57, and Ag = 107·88.*

STEP I

Find the eq. wts. of copper and silver.

$$\text{Eq. wt.} = \frac{\text{At. wt.}}{\text{Valency}}$$

$$\therefore \quad \text{Eq. wt. of copper} = \frac{63 \cdot 57}{2} = 31 \cdot 78$$

$$\text{and eq. wt. of Ag} = \frac{107 \cdot 88}{1} = 107 \cdot 88.$$

Equivalent Weights

STEP II

Find the weight of silver.

Applying Faraday's second law of electrolysis we have

$$\frac{\text{Wt. of copper}}{\text{Wt. of silver}} = \frac{\text{eq. wt. of copper}}{\text{eq. wt. of silver}}$$

Substituting the values, $\dfrac{0.35}{x} = \dfrac{31.78}{107.88}$

$$\therefore \quad x = \frac{0.35 \times 107.88}{31.78} = 1.188 \text{ g}$$

Thus wt. of copper deposited $= 1.188$ g

9. Miscellaneous

Example 16. *1.65 g of a metal nitrate gave 1.115 g of its oxide. Find the equivalent weight of the metal.*

We are given :

Wt. of metal nitrate $= 1.65$ g

Wt. of metal oxide $= 1.115$ g

Applying the law of equivalents,

$$\frac{\text{Wt. of metal nitrate}}{\text{Wt. of metal oxide}} = \frac{\text{eq. wt. of metal nitrate}}{\text{eq. wt. of metal oxide}}$$

$$= \frac{\text{eq. wt. of the metal} + \text{eq. wt. of NO}_3\text{-radical}}{\text{eq. wt. of metal} + \text{eq. wt. of O}_2\text{-radical}}$$

or $\quad \dfrac{1.65}{1.115} = \dfrac{E + 62}{E + 8}$

or $\quad 1.65(E + 8) = 1.115(E + 62)$

or $\quad\quad\quad\quad E = 104.5$

$\therefore$ Equivalent weight of the metal $= 104.5$.

Example 17 *1.5 g of a metal carbonate was heated and 400.87 ml of CO_2 was evolved at 27°C and 700 mm pressure. Calculate the equivalent weight of the metal.*

STEP I

Calculate the volume of CO_2 at S.T.P.

Using the equation

$$\frac{P_1 V_1}{T_1} = \frac{P_2 V_2}{T_2}.$$

$$P_1 = 700 \text{ mm} \qquad P_2 = 760 \text{ mm}$$
$$V_1 = 400\cdot87 \text{ ml} \qquad V_2 = ?$$
$$T_1 = 300° \text{ Abs} \qquad T_2 = 273° \text{ Abs}$$

$$\therefore \quad \frac{700 \times 400\cdot87}{300} = \frac{760 \times V_2}{273}$$

or $$V_2 = 336 \text{ ml}$$

STEP II

Find the weight of CO_2.

Now 22400 ml of CO_2 weigh at S.T.P. = 44 g

1 ml of CO_2 weighs at S.T.P. = $\dfrac{44}{22400}$

336 ml of CO_2 weigh at S.T.P.

$$= \frac{44}{22400} \times 336 = 0\cdot66 \text{ g}$$

STEP III

Calculate the eq. wt. of the metal.

Applying the law of equivalents

$$\frac{\text{Wt. of metal carbonate}}{\text{Wt. of } CO_2} = \frac{\text{Eq. wt. of metal carbonate}}{\text{Eq. wt. of } CO_2}$$

$$\frac{1\cdot5}{0\cdot66} = \frac{E+30}{22}$$

or $$E = 20$$

EQUIVALENT WEIGHT OF COMPOUNDS

Equivalent Weight of Acids. Since all acids contain replaceable hydrogen atoms. the *gram equivalent weight of an acid is that weight of it which contains one gram equivalent weight of hydrogen (replaceable).*

One mole of HCl contains one replaceable hydrogen atom only, hence the gram equivalent weight of HCl is

$$1 + 35\cdot5 = 36\cdot5 \text{ g}$$

One mole of the sulphuric acid contains two replaceable hydrogen atoms thus, its gram equivalent is

$$\tfrac{1}{2}(2+32+64) = 49$$

Equivalent Weights

It should however be noted that the nature of the reaction in which H_2SO_4 takes part will determine its exact equivalent weight. For example, in the reaction

$$Na_2CO_3 + H_2SO_4 \longrightarrow NaHSO_4 + H_2O + CO_2,$$

the equivalent weight of H_2SO_4 is 98 while in the reaction

$$Na_2CO_3 + H_2SO_4 \longrightarrow Na_2SO_4 + H_2O + CO_2,$$

the equivalent weight of H_2SO_4 is 49. In the same way, the equivalent weight of H_3PO_4 may be one mole, half mole, or one-third mole depending upon the number of hydrogen atoms actually taking part in the reaction (*i.e.*, 1, 2 or 3).

Equivalent Weight of Bases. All bases contain 'OH' (hydroxyl) groups. The equivalent weight of a base depends upon the number of replaceable hydroxyl groups present in the molecule of the base. Thus **one gram equivalent weight of a base** *is that weight of the base which contains 1 gram equivalent weight of the hydroxyl ion (or radical) i.e., 17·008.* The equivalent weight of NaOH is $23+16+1=40$, that of KOH is $39+16+1=56$. The equivalent weight of $Ca(OH)_2$ in the reaction.

$$Ca(OH)_2 + HCl \longrightarrow Ca(OH)Cl + H_2O$$

will be equal to its molecular weight *i.e.*, 74 but in the reaction

$$Ca(OH)_2 + 2HCl \longrightarrow CaCl_2 + 2H_2O$$

it is equal to half mole *i.e.*, $\frac{74}{2} = 37$.

In the same way, the equivalent weight of $Fe(OH)_3$ may be its one mole, $\frac{1}{2}$ mole or $\frac{1}{3}$ mole depending upon the number of hydroxyl groups actually taking part in the reaction.

Equivalent Weight of Salts. The equivalent weight of a simple salt (non-oxidising and reducing agent) is *that weight of it which contains one gram equivalent of the metal or the radical.* For example, the equivalent weight of KCl is 74·5 since it contains one gram equivalent weight of potassium (39) or chlorine (35·5). Similarly, the equivalent weight of aluminium chloride ($AlCl_3$) is that weight of $AlCl_3$ which contains 1 gram equivalent of Al (9) or 1 gram equivalent of chlorine (35·5). Now the molecular weight of $AlCl_3$ is 133·5 ($=27+106·5$), the equivalent weight of $AlCl_3$ will be $\frac{1}{3}$ (133·5) or 44·5.

Equivalent Weight of Salts on the Concept of Valency. The valency of elements, hydrogen, chlorine or oxygen which are reference standards in the determination of equivalent weights of other elements or compounds, is also helpful in finding the equivalent weight of substances. For example, we take the case of $CaCl_2$. The molecular weight of $CaCl_2$ is $40+71=111$. One atom of calcium is present per mole of $CaCl_2$. The valency of calcium is 2, hence the equivalent weight of $CaCl_2$ will be $\frac{1}{2}(111) = 55.5$. **Thus the equivalent weight of a compound on the concept of valency, is its molecular weight divided by the total positive or negative valency.** In case of calcium phosphate, $Ca_3(PO_4)_2$, the molecular weight is $3 \times 40 + 2 \times 31 + 8 \times 16 = 310$.

Total positive valency (*i.e.*, valency of 3 calcium atoms) $3 \times 2 = 6$ or total negative valency (two phosphate radicals) is 2×3 *i.e.*, 6.

Thus equivalent weight of $Ca_3(PO_4)_2 = \dfrac{310}{6}$

$= 51.66$

Example 18. *Find the equivalent weight of*

(i) HNO_3 (ii) HBr

assuming complete neutralization.

(i) **HNO_3**

Molecular weight $= 1 + 14 + 48 = 63$

Now one mole of HNO_3 contains only one gram atom of hydrogen.

Hence 1 gram equivalent of $HNO_3 = 63$ g

(ii) **HBr**

Molecular weight $= 1 + 80 = 81$

Now one mole of HBr contains only one gram-atom of hydrogen.

Hence 1 gram equivalent of $HBr = 81$ g

Example 19. *Calculate the equivalent weights of*

(i) $Al(OH)_3$ (ii) $NaOH$

(i) **$Al(OH)_3$**

Equivalent Weights

1 gram equivalent of $Al(OH)_3$ will be that quantity which contains one gram equivalent of replaceable OH.

Now molecular weight (or Formula weight) $= 27+51$
$$= 78$$

Since $Al(OH)_3$ has three replaceable OH^-

Its gram equivalent weight $= \dfrac{\text{Molecular weight}}{3}$

$$= \dfrac{78}{3} = 26 \cdot 0 \text{ g}$$

(ii) NaOH

Molecular weight $= 23+16+1 = 40$

No. of replaceable OH^- ions $= 1$

$\therefore$ equivalent weight NaOH $= \dfrac{40}{1} = 40$

Example 20. *Calculate the equivalent weights of*

(i) $AlPO_4$ (ii) $(NH_4)_2SO_4$

(iii) $FeCl_3 \cdot 6H_2O$

(i) $AlPO_4$

STEP I

Find the molecular weight of $AlPO_4$.

Molecular weight of $AlPO_4 = 1 \times Al + 1 + P + 4 \times 0$
$$= 27 + 31 + 64$$
$$= 122$$

STEP II

Find the total positive valency.

Since there is only one Al^{3+}.

$\therefore$ Total positive valency $= 1 \times$ valency of Al

$$1 \times 3 = 3$$

STEP III

Find the equivalent weight

Equivalent weight $= \dfrac{122}{3}$

$$= 40 \cdot 66$$

(ii) **$(NH_4)_2SO_4$**

$$\text{Molecular weight} = 2 \times 18 + 32 + 64$$
$$= 132$$

Total positive valency of $2NH_4^+$
$$= 2 \times \text{Valency of } NH_4$$
$$= 2 \times 1 = 2$$

$\therefore \quad$ Éq. wt. $= \dfrac{\text{Mol. wt.}}{\text{Total positive valency}}$

$$= \dfrac{132}{2} = 66$$

(iii) **$FeCl_3 \cdot 6H_2O$**

$$\text{Molecular weight} = 1 \times 56 + 3 \times 35 \cdot 5 + 6 \times 18$$
$$= 56 + 106 \cdot 5 + 108$$
$$= 270 \cdot 5$$

Total positive valency (of one Fe^{3+}) $= 1 \times 3 = 3$

$$\text{Eq. wt.} = \dfrac{207 \cdot 5}{3} = 69 \cdot 116$$

Equivalent Weight of an Oxidising Agent. The equivalent weight of an oxidising or a reducing agent is equal to the molecular weight of the compound of a particular element divided by the change in the oxidation state taking part in a particular reaction.

Thus, $\quad$ eq. wt. $= \dfrac{\text{Molecular weight}}{\text{Total change in oxidation state}}$

Example 21. *Find the oxidation state of N, Cl, I, S in the compounds NH_3, HCl, HI, H_2S, H_2SO_4 and HNO_3. Given that the oxidation state of hydrogen is $+1$ and oxygen -2.*

The total oxidation state of a compound is always zero.

Thus in NH_3 the total oxidation state of hydrogen
$$= +1 \times 3 = +3$$

(The oxidation state of one hydrogen is $+1$, so that of three hydrogen atoms it is $+3$).

Since the oxidation state of ammonia (being a compound) is zero, the oxidation state of N is $=-3$.

Similarly, in **HCl**, since the total oxidation state of hydrogen $=+1$.

∴ The oxidation state of $Cl = -1$.

in **HI**, the total oxidation state of hydrogen $=+1$.

∴ The oxidation state of $I = -1$.

in **H₂S**, the total oxidation state of hydrogen $=+2$.

∴ The oxidation state of $S = -2$.

in **H₂SO₄**, the total oxidation state of $H_2 = +2$

the total oxidation state $O_4 = -2 \times 4 = -8$

∴ The oxidation state of $S = +6$

in **HNO₃**, the total oxidation state of hydrogen $=+1$

the total oxidation state of oxygen $= -2 \times 3 = -6$

∴ The oxidation state of $N = +5$.

Example 22. *Calculate the equivalent weight of $K_2Cr_2O_7$ in the reaction*

$$K_2Cr_2O_7 + 4H_2SO_4 \longrightarrow K_2SO_4 + Cr_2(SO_4)_3 + 4H_2O + 3O.$$

Before finding the equivalent weight of an oxidising agent, we have to see as to which element undergoes a change in the oxidation state during the course of reaction. Here, in the above reaction, element chromium undergoes a change in the oxidation state.

STEP I

Find the molecular weight of $K_2Cr_2O_7$

$$= 2 \times 39 + 2 \times 52 + 7 \times 16$$
$$= 78 + 104 + 112$$
$$= 294.$$

STEP II

Find the change in the oxidation state (no.)

Oxidation state of 2Cr in $K_2Cr_2O_7$ is found by taking into account the oxidation numbers of the other elements.

8—NPC

Oxidation no. of $2K = +2$
Oxidation no. of $7O = -14$
∴ Oxidation no. of $2Cr = +12$
or   Oxidation no. of $Cr = +6$.

Hence oxidation state of Cr in $K_2Cr_2O_7$ is $+6$

Oxidation no. of $3(SO_4) = 3 \times -2$ (the oxid. no. of SO_4^{2-})
$= -6$.

∴ Oxidation no. of $2Cr = +6$
or   Oxidation no. of $Cr = +3$.

Since 2 Cr take part in the reaction throughout, the total the change in oxidation state $= 12 - 6 = 6$.

STEP III

Find the equivalent weight of $K_2C_2rO_7$

$$\text{Eq. wt.} = \frac{\text{Mol. wt.}}{\text{Total change in oxid. number}}$$

$$= \frac{294}{6} = 49.$$

Example 23. *Find the equivalent weight of* $KMnO_4$ *in the reaction*

$$2KMnO_4 + 3H_2SO_4 \longrightarrow K_2SO_4 + 2MnSO_4 + 3H_2O + 5O$$

STEP I

Find the Mol. weight of $KMnO_4$

$$= 39 + 55 + 64$$
$$= 158.$$

STEP II

Find the total change in oxidation number of (state).

Oxidation number of Mn undergoes a change in this reaction.

The oxidation state of Mn in $KMnO_4$ is

$$K = +1$$
$$Mn = ?$$
$$O_4 = -8$$

Equivalent Weights

∴ The oxidation state of $Mn = +7$.

Oxidation state of Mn in $MnSO_4$ is

$$Mn = ?$$
$$SO_4 = -2$$

∴ Oxidation state of $Mn = +2$.

Change in oxidation state $= 7 - 2 = 5$.

Step III

Find the equivalent weight of $KMnO_4$.

$$\text{Eq. wt.} = \frac{\text{Mol. wt.}}{\text{Total change in oxidation state}}$$

$$= \frac{158}{5} = 31 \cdot 6.$$

END-OF-CHAPTER PROBLEMS

Hydrogen Displacement Method

1. When 1·25 g of zinc is treated with excess H_2SO_4. 0·428 litre hydrogen is evolved at N.T.P. Find the equivalent weight of zinc. [Ans. 32·71]

2. A quantity of aluminium weighing 0·250 g on treatment with dil. H_2SO_4 gave 310 ml of H_2 at N.T.P. Find the eq. wt. of Al. [Ans. 9·03]

3. 0·650 g on zinc metal on treatment with dil. H_2SO_4 gave 246 ml of H_2 at 27°C and 760 mm pressure. Calculate the eq. wt. of the zinc metal. [Ans. 32·26]

4. 0·309 of a metal on treatment with dil. H_2SO_4 gave 246 ml of H_2 at 17°C and 775 mm pressure. Calculate the eq. wt. of the metal. (Aq. tension at 17°C = 14·4 mm.) [Ans. 23·3]

5. 1·308 g of a metal on treament with dil. H_2SO_4 was found to give rise to 512·6 ml of hydrogen at 27°C and 756·5 mm. Find out the equivalent weight of the metal (Aq. tension at 27°C is 26·5 mm.) [Ans. 32·68]

6. 0·225 g of zinc metal when treated with dil. H_2SO_4 gave 82·3 ml of H_2 at 23°C and mm pressure. Calculate the eq. weight of zinc (Aq. tension at 23°C is 21 mm). [Ans. 32·67]

7. 0·2178 g of magnesium metal when dissolved in hydrochloric acid gave 218·2 ml of hydrogen collected over water at 17°C and 754·5 mm pressure. Calculate the eq. weight of Magnesium (Aq. tension at 17°C is 14·5 mm). [Ans. 12·2]

8. 0·2 g of a metal gave on treatment with dilute mineral acid 68·4 ml of hydrogen measured at N.T.P. Calculate the equivalent weight of the metal. [**Ans.** 32·75]

9. 0·654 g of metal on treatment with dil. sulphuric acid evolved 256·3 cc moist hydrogen at 27°C and 756·5 mm pressure. Find out the equivalent weight of the metal (Aq. tension at 27°C=26·6 mm).
(*Haryana Board Hr. Sec., 1978*)
[**Ans.** 32·53]

10. How many hydronium ions (hydrogen ions) will be furnished by two gram-equivalents of hydrochloric acid ?
(*Guru Nanak Dev Pre-Univ., 1978*)
[**Ans.** 12.046×10^{23}]

Oxide Formation Method

11. 2·00 g of tin on treatment with concentrated nitric acid and subsequent ignition gave 2·539 g of its oxide. Calculate the equivalent weight of the metal. [**Ans.** 29·70]

12. 1·62 g of zinc when dissolved in nitric acid gave zinc nitrate The nitrate when strongly heated left a residue of zinc oxide which accurately weighed 2·02 g. Find the eq. wt. of zinc. [**Ans.** 32·4]

13. 1·201 g of zinc gave 1·497 g of zinc oxide on treatment with nitric acid and subsequent ignition. In a second experiment 0·543 g of zinc precipitated 0·527 g of copper from a solution of copper sulphate. Calculate the eq. wt. of copper and zinc, assuming that of oxygen to be 8. [**Ans.** zinc=32·46, Cu=31·51]

14. 2·043 g of zinc combines with 0·5 g of oxygen. What is the equivalent weight of zinc ? [**Ans.** 32·7]

15. Determine the equivalent weight of tin in :
(*a*) the oxide which contains 78·6% tin and
(*b*) the oxide which contains 88·1% tin.
[**Ans.** (*a*) = 29·6 ; (*b*) = 59·3]

16. A metal oxide contains 59·2% metal. What is the equivalent weight of the metal ? [**Ans.** 11·6]

17. 1·986 gm. of copper produces 2·47 gm of copper oxide. Find the equivalent weight of copper.
(*Kurukshetra Pre-Univ., 1973*)
[**Ans.** 32·8]

18. 3·0 g of a carbonate of a metal on strong heating gave 1·69 g of its oxide. Calculate the equivalent weight of the metal.
(*Himachal Pre-Univ., 1981*)
[**Ans.** 25]

19. 3·54 gm of a metal were dissolved in excess of concentrated HNO_3 and the nitrate thus formed was ignited carefully. The weight of the oxide formed was 4·14 gm. Calculate the equivalent weight of the metal. (*Kurukshetra Pre-Univ., 1982*)
[**Ans.** 31·46]

Equivalent Weights

Reduction of Oxide Method

20. What is the equivalent wt. of iron in each of the following oxides:

(i) In one oxide Fe is 77·727%
(ii) In second oxide Fe is 72·359%
(iii) In the third oxide Fe is 70·00%

[**Ans.** 27·9, 29·94, 18·66]

21. 2·00 g of a metallic oxide yielded 1·748 g of pure metal. Calculate the equivalent weight of the metal. [**Ans.** 55·05]

22. 2·54 g of the oxide of a metal was heated in the atmosphere of hydrogen when 0·607 g of water was formed in the reaction. Calculate the equivalent weight of the metal. [**Ans.** 29·63]

23. A compound of nitrogen and oxygen contains 63·6% of nitrogen and 36·4% of oxygen. What is the equivalent weight of nitrogen.

(*Calicut Pre-Degree.*, 1980)
[**Ans.** 13·98]

24. 3·50 g of a metal was dissolved in excess of nitric acid and the nitrate thus formed was ignited and formed 4·80 g of metal oxide. Find the equivalent weight of the metal.

(*Punjabi Pre-Univ.*, 1980)
[**Ans.** 21·54]

Formation of Chloride Method

25. 0·5302 g of a metal yielded 0·7052 g of its chloride when heated in an atmosphere of chlorine gas. Calculate the equivalent weight of the metal? [**Ans.** 107·6]

26. The analysis of the chloride of a metal showed it to contain 24·8% chlorine, the rest being the metal. What is the equivalent weight of the metal? [**Ans.** 109·1]

27. The chloride of a non-metal contains 77·4% chlorine. Find the eq. wt. of the non-metal. [**Ans.** 10·33]

28. 0·5 g of a metal was dissolved in nitric acid, on adding HCl solution, the chloride of this metal obtained was 0·665 g. Find the eq. wt. of the metal. [**Ans.** 107·6]

29. The chloride of a metal contained 65·61% chlorine and 34·39% metal. Find the equivalent weight of the metal.

[**Ans.** 18·61]

30. The chloride of a metal contains 65% chlorine. Find out its Equivalent Weight. (*Himachal Pre-Univ.*, 1978)
[**Ans.** 19·11]

31. 1·11 gm of the chloride of a metal dissolved in water were treated with excess of silver nitrate solution. The weight of silver chloride was found to be 2·87 gm. Calculate the equivalent weight of the metal. (Eq. Wt. of Ag = 108; Cl = 35·5).

(*Punjabi Pre-Univ.*, 1982)
[**Ans.** 55·5]

Double Decomposition Method

32. 1·17 g of metal chloride were dissolved in water and treated with excess of $AgNO_3$ solution. The weight of the precipitated $AgCl$ was found to be 2·87 g. Calculate the eq. weight of the metal.

[Ans. 20·00]

33. 1 g of metal chloride solution when treated with silver nitrate gave 0·687 g of dry $AgCl$ precipitate. What is the equivalent weight of the metal when that of Ag and chlorine are 108 and 35.5 ?

[Ans. 68·5]

34. 0·5 g of the chloride of a metal was dissolved in water and treated with excess of $AgNO_3$ solution. 1·225 g of $AgCl$ was obtained. Find the eq. wt. of the metal.

[Ans. 23]

35. 1 g of sodium sulphate was dissolved in water and treated with barium chloride. 1·94 g of $BaSO_4$ was obtained. Given that the equivalent weight of sodium and sulphate are 23 and 48 respectively, find that of the Barium.

[Ans. 68·5]

36. 1·03 g of anhydrous barium chloride was dissolved in water and treated with excess of sulphuric acid. The weight of dry Barium sulphate obtained was 1·16 g. Find the eq. wt. of barium metal.

[Ans. 68·5]

37. 1·225 g of a metal sulphate were produced as a precipitate by the addition of excess of sodium sulphate solution to the solution containing 1 g of the chloride of the metal. Calculate the equivalent weight of the metal.

[Ans. 20·05]

Metal Displacement Method

38. 2·00 g of magnesium metal was suspended in aluminium sulphate solution. 1·50 g of aluminium metal was precipitated. If the eq. wt. of magnesium is 12, find that of aluminium.

[Ans. 9·0]

39. 1·30 g of zinc metal displaced 1·272 g of copper from a solution of copper sulphate. If the eq. wt. of zinc be 32·5, find that of copper.

[Ans. 31.8]

40. 5 g of calcium metal displaced 3 g of magnesium from the solution of magnesium chloride. If the eq. wt. of magnesium is 12, find that of calcium.

[Ans. 20]

41. 0·70 g of iron precipitated 0·795 g of copper from copper sulphate solution. Find the equivalent weight of copper if that of iron is 28·0.

[Ans. 31·5]

42. 1·00 g magnesium displaced 0·75 g of aluminium from a solution of aluminium chloride. 2 g calcium displaced 0·90 g of aluminium. If the equivalent weight of calcium is 20, find that of magnesium.

[Ans. 12]

43. 1·201 g of zinc on treatment with nitric acid and subsequent ignition gave 1·497 g of zinc oxide. In the second experiment 0·543 g of zinc precipitated 0·527 g of copper from copper sulphate

solution. Calculate the equivalent weight of copper and zinc assuming that of oxygen to be 8. *(Punjabi Pre-Univ., 1978)*
[**Ans.** Eq. wt. of Zn = 32·45]
Eq. wt. of Cu = 31·49]

Faraday's Electrolytic Method

44. How many coulombs per hour pass through an electrolytic cell which uses a current of 2·5 amperes ? How many gram-equivalents of silver will be deposited by this current ?
[**Ans.** *(i)* 900 coulombs ;
(ii) 0·09 gm eqts.]

45. What current strength is required to pass 96·500 coulombs per hour through a cell ? How much aluminium will be deposited by this current in one hour ? [**Ans.** 26·8 amp., 9·00 gm]

46. When a current strength of 10 amperes is passed through a solution of zinc sulphate for 30 minutes it deposits 6·099 g of zinc. Find the eq. wt. of zinc metal. [**Ans.** 32·7]

47. The eletro-chemical constant of magnesium is 0·000124, find the equivalent weight of magnesium. [**Ans.** 12·00]

48. Find the weight of copper deposited from copper sulphate solution by current of 0·25 ampere flowing for one hour.
[**Ans.** 0·2936 g]

49. In a cell fitted with platinum electrodes 0·5 ampere current is passed for 20 minutes through copper sulphate solution. Atomic weight of copper is 63·5.

(i) What weight of Cu would be liberated ?

(ii) How much oxygen in c.c. would be produced at N.T.P. ?
(North Bengal Pre-Univ., 1973)
[**Ans.** *(i)* 0·198 gm ; *(ii)* 35·00 ml nearly]

50. An electric current is passed through a solution of copper sulphate and a solution of zinc sulphate connected in series. If in a given time the weight of copper deposited is one gm, what would be the weight of zinc deposited in the same time ? (eq. wt. of Cu=31·8 ; eq. wt.=32·7) *(Shivaji Pre-Univ., 1975)*
[**Ans.** 1·028 gm]

51. The same current has passed through acidulated water and a solution of copper sulphate. It liberated 203 ml of hydrogen measured at N.T.P. and deposited 0·578 gm of copper. Calculate the equivalent weight of copper, that of hydrogen being one.
(Guru Nanak Dev Pre-Univ., 1977)
[**Ans.** 31·8]

52. A current of 500 milliamperes was passed through a solution of $AgNO_3$ for 10 minutes. What weight of silver will be deposited on the cathode if the electrochemical equivalent of silver is 0·00111
(Guru Nanak Dev Pre-Univ., 1981)
[**Ans.** 0.333 g]

53. The same current was passed through acidulated water and a solution of $CuSO_4$. It liberated 203 ml of hydrogen measured at NTP and deposited 0·578 g of copper. Calculate the equivalent weight of copper.

(Punjab Pre-Univ., 1981)
[**Ans.** 31·636]

Miscellaneous

54. 2·52 g of a pure carbonate were heated strongly and 1·2 g of the metal oxide was obtained. Calculate the equivalent weight of the metal.
[**Ans.** 12]

55. 0·444 g of a metal on treatment with dil. H_2SO_4 evolved 456 ml of moist hydrogen at 17°C and 740 mm pressure. Calculate the equivalent weight of the metal. (Aq. tension at 17°C is 15 mm).
[**Ans.** 12·14]

56. 1·659 g of a metal carbonate on heating gave 542·8 ml of CO_2 at 27°C and 700 mm pressure. Find the eq. wt. of the metal.
[**Ans.** 20·0]

57. 0·75 g of a metal iodide gave 0·292 g of the metal chloride with chlorine gas. Calculate the eq. wt. of the metal. [**Ans.** 23·00]

58. 0·420 g of a monovalent metal carbonate (M_2CO_3) neutralizes exactly 100 ml of decinormal sulphuric acid. Calculate equivalent weight of the metal M. (C=12. O=16)

(Kurukshetra Pre-Univ., 1981)
[**Ans.** 12]

Equivalent Weight of Compounds

59. Calculate the gram-equivalent weight of

(i) LiOH (ii) KOH (iii) $Ca(OH)_2$ (iv) $Al(OH)_3$ (v) $Fe(OH)_3$
[**Ans.** (i) 24 g (ii) 56 g (iii) 37 g (iv) 26 g (v) 35·66 g]

60. Calculate the gram equivalent weight of

(i) HCl (ii) HNO_3 (iii) H_3PO_4 (iv) H_2CO_3
[**Ans.** (i) 36·5 g (ii) 63 g (iii) 32·66 g (iv) 31]

61. Find the equivalent weight of the metal in the following compounds ;

(i) Na_2SO_4 (ii) Hg_2Cl_2 (iii) $BaCl_2$ (iv) HgO
[**Ans.** (i) 23 (ii) 200·6 (iii) 68·5 (iv) 100·3]

62. What would be the equivalent weight of iron when it is involved in the following reactions :

(i) $2FeSO_4 + H_2SO_4 + \frac{1}{2}O_2 \longrightarrow Fe_2(SO_4)_3 + H_2O$
(ii) $\qquad 2Fe + 3Cl_2 \longrightarrow 2FeCl_3$

(Punjab Pre-Univ., 1974)
[**Ans.** (i) 56 (ii) 18·6]

63. 1·2 g of metal NaCl was dissolved in water and solution was treated with excess of $AgNO_3$. The weight of AgCl formed after drying and weighing was found to be 2·94 g. Calculate the equivalent

weight of sodium when equivalent weights of silver and chlorine are 108 and 35·5 respectively. *(Kurukshetra Pre-Univ., 1976)*
[**Ans.** 23·07]

64. 0·205 g of a metal on treatment with an acid gave 10·66 ml of moist hydrogen collected at 755 mm pressure and 17°C. Calculate the equivalent weight of the metal.

(Aqueous tension at 17°C=14·4 mm)
(Punjabi Pre-Univ., 1974)
[**Ans.** 23·48]

65. Calculate the equiavlent weight of magnesium if 0·724 g of it displaces 6·450 g of silver from its salt. (Eq. wt. of Ag=107·88).
(Himachal Pre-Univ., 1976)
[**Ans.** 12·10]

66. The same current was passed through acidulated water and a solution of copper sulphate. It liberated 205 ml of hydrogen measured at NTP and deposited 0·578 g of copper. Calculate the equivalent weight of copper, that of hydrogen being one.
(Guru Nanak Dev Pre-Univ., 1975)
[**Ans.** 31·32]

67. A current of 0·5 ampere passes through a solution of $AgNO_3$ for 6 minutes. Find the weight of silver deposited. (At. wt. of Ag=108) *(Gauhati Pre-Univ., 1975)*
[**Ans.** 0·2015]

68. How many amperes of current are required to liberate 2·24 litres of chloride at NTP in one hour during electrolysis of sodium chloride solution. *(Maharaja Sayaji Rao Pre-Univ., 1976)*
[**Ans.** 2·68]

69. What is the Equivalent Weight of $SnCl_2$ in the following reaction :

$$SnCl_2 + Cl_2 \longrightarrow SnCl_4$$

(At. wts. Sn=119, Cl=35·5)

(Himachal Pre-Univ., 1978)
[**Ans.** 85]

10

Atomic Weights

Atomic weight *of an element is the average relative weight of its atoms as compared to an atom of carbon as 12.* The choice of the standard for referring the atomic weights is arbitrary. For a long time, the reference standerd was hydrogen, then came oxygen and presently the most common variety of carbon (at. wt. 12) is the reference standard. The disadvantage of hydrogen and oxygen as standards is that they do not combine with as many elements as carbon does and hence the utility of carbon. Since we find the atomic weights of elements relative to one standard or the other, these atomic weights are relative weights. Now when we say that the atomic weight of Mg is 24, it means that the atom of magnesium is roughly 24 times heavier than hydrogen or $\frac{24}{16}$ times heavier than oxygen or $\frac{24}{12}$ times heavier than carbon.

GRAM-ATOM OR GRAM-ATOMIC WEIGHT

Gram-atomic weight of an element is the atomic weight of an element expressed in grams. Thus one gram-atom of hydrogen means 1·008 g of hydrogen, one gram-atom of carbon means 12 g of carbon. When we say two gram-atoms of chlorine we mean $2 \times 35·5$ g of chlorine. According to Avogadro's number one gram-atom of any element contains $6·02 \times 10^{23}$ atoms of that element in it. Thus one gram-atom of sodium

(23 g) will contain 6.02×10^{23} atoms of sodium or we can say that 6.02×10^{23} atoms of sodium weigh equal to its gram-atomic weight. From this relation it is possible to find the actual weight of an element. Since the atomic weight of hydrogen is taken to be 1.008, it means

that 6.02×10^{23} atoms of hydrogen weight
$$= 1.008 \text{ g}$$

$\therefore$ 1 atom of hydrogen will weigh $= \dfrac{1.008}{6.02 \times 10^{23}}$

$$= 1.66 \times 10^{-24} \text{ g}$$

Realtion between Atomic Weight and Equivalent Weight of an element

Let us have an element whose atomic weight $= A$
let the equivalent weight of this element $= E$
and the valency of the element $= v$.

From the definition of valency, we can say that v atoms of the element combine with 1 atom of the element since the atomic weight of 1 atom of hydrogen is taken to be approx. 1.

$\therefore$ the weight of v atoms of hydrogen $\times v \times 1 = v$

thus v parts by weight of hydrogen combine with element
$$= A \text{ parts (At. wt.)}$$

$\therefore$ 1 part by weight of hydrogen combines with element
$$= \dfrac{A}{v} \text{ parts}$$

and by definition the equivalent weight of a substance is the number of parts by weight of the element which combine with 1 part by weight of hydrogen.

Thus $\dfrac{A}{v}$ parts is the equivalent weight of this element.

Hence $\qquad E = \dfrac{A}{v}$

or $\qquad E \times v = A$.

Thus Atomic weight $=$ Equivalent weight $\times$ valency.

Example 1. *The equivalent weight of an element is 9, its valency is 2. Calculate the atomic weight.*

Applying the relation

$$\text{At. wt.} = \text{Valency} \times \text{eq. wt.}$$

and substituting the values, we have

$$\text{At. wt.} = 3 \times 9 = 27$$

Example 2. *A 0·627 g oxide of a divalent metal on reduction gave 0·500 g of the metal. Find the atomic weight of the metal.*

Step I

Find the equivalent weight of the metal.

Wt. of the oxide = 0·627 g

Wt. of the metal = 0·500 g

Wt. of oxygen combining with 0·500 g metal

$$= 0.127 \text{ g}$$

Now 0·127 g oxygen combines with metal = 0·500 g

$$1 \text{ g oxygen combines with metal} = \frac{0.500}{0.127} \text{ g}$$

$$\therefore \quad 8 \text{ g oxygen combines with metal} = \frac{0.500}{0.127} \times 8 \text{ g}$$

Thus equivalent wt. is = 31·5

Step II

Find the atomic weight.

Applying the relation

$$\text{At. wt.} = \text{Valency} \times \text{Eq. wt.}$$

and substituting the values, we have

$$\text{At. wt.} = 2 \times 31.5 = 63$$

(Since the metal is *divalent*, its valency is 2).

METHODS FOR FINDING ATOMIC WEIGHTS

There are a number of methods available for finding the atomic weights of elements. We shall presently take up **Dulong and Petit's method** for finding atomic weights.

Dulong and Petit's method

This method is based on Dulong and Petit's law which states that *"the product of the specific heat and atomic weight of an element (called atomic heat) is 6·4 approximately."*

Mathematically,

Atomic weight × specific heat = 6·4 approx. (at 22°C).

Since one gram-atom of every element has $6·02 \times 10^{23}$ atoms in it, the value of 6·4 being for all such atoms of the element, is a constant quantity. This means that if we know the specific heat of an element we can find its atomic weight as well. Since the value 6·4 is approximate, the atomic weight thus found out will also be approximate.

Determination of Exact Atomic Weight

The following steps are involved in the determination of the exact atomic weight of an element.

(1) Find the approximate atomic weight of the elements by applying the relation

At. wt. × Sp. heat = 6·4 approx.

(2) Find the equivalent weight of the element by a suitable method or from the data given.

(3) Find the valency of the element by applying the relatsonship

At. wt. = Valency × Equivalent weight.

(4) The valency so obtained is usually a fractional number. Make it nearest whole number if not so already.

(5) Find the exact atomic weight by substituting the values of whole number valency and equivalent weight in the relation.

At. wt. = Valency × Eq. wt.

Example 3. *The oxide of an element is found to contain 20% oxygen. The specific heat of the element is 0.068. Calculate the equivalent weight, valency and atomic weight of the element.*

STEP I

Find the eq. wt. of the metal.

Wt. of oxide = 100 g

Wt. of oxygen = 20 g

∴ Wt. of the element = 80 g

Now, 20 g oxygen combines with element = 80 g

∴ 1 g oxygen combines with element = $\frac{80}{20}$ g

and 8 g oxygen combines with element = $\frac{80}{20} \times 8$ g

Thus equivalent weight is = 32

STEP II

Find the approx. At. wt.

Applying the relation

Approx. at. wt. × sp. heat = 6·4

and subtituting the values,

$$\text{Approx. at. wt} = \frac{6 \cdot 4}{0 \cdot 068} = 94 \cdot 1$$

STEP III

Find the valency.

$$\text{At. wt.} = \text{Valency} \times \text{Eq. wt.}$$
$$94 \cdot 1 = \text{Valency} \times 32$$

or Valency = $\frac{94 \cdot 1}{32}$ = 2·95 = 3 (whole number)

STEP IV

Find the exact At. wt.

Exact Atomic weight = Valency × Eq. wt. = 3 × 32 = 96.

Example 4. *1·0 g of a metal yielded 1·328 g of its chloride. The specific heat of the metal is 0·059. Find the atomic weight of the metal.*

STEP I

Find the Eq. wt. of the metal.

Wt. of metal = 1·0 g

Wt. of metal chloride = 1·328 g

Atomic Weights

∴ Wt. of chloride which combines with **1·0 g** metal
$$= 1·328 - 1·0 = 0·328 \text{ g}$$

Now 0·328 g of chlorine combines with metal $= 1·0$ g

1 g of chlorine combines with metal $= \dfrac{1}{0·328}$

35·5 g of chlorine combines with metal
$$= \dfrac{1}{0·328} \times 35·5 = 108·2 \text{ (eq. wt.)}$$

Step II

Find the approx. At. wt.

$$\text{Approx. At. wt.} = \dfrac{6·4}{\text{Sp. heat}}$$
$$= \dfrac{6·4}{0·059} = 108·4$$

Step III

Find the Valency.

Applying the relation
$$\text{At. wt.} = \text{Valency} \times \text{Eq. wt.}$$
$$\text{Valency} = \dfrac{\text{At. wt.}}{\text{Eq. wt.}} = \dfrac{108·2}{108·4}$$
$$= 1 \text{ (Whole number)}$$

Step IV

Find the exact atomic weight.

$$\text{Exact at. wt.} = \text{Whole no. Valency} \times \text{Eq. wt.}$$
$$= 1 \times 108·2 = 108·2.$$

Example 5. *0·44 g of a metal with sepecific heat 0·107 when dissolved in hydrochloric acid produced 117 ml of hydrogen at 10°C and 759 mm pressure. Calculate the exact atomic weight of the element.*

Step I

Find the equivalent weight of the metal

Vol. of hydrogen $(v_1) = 177$ ml

$$T_1 = 273 + 10 = 283°\text{Abs}$$
$$P_1 = 759 \text{ mm}$$

Vol. at S.T.P. is found by the equation

$$\frac{P_1 V_1}{T_1} = \frac{P_2 V_2}{T_2}$$

$$\frac{759 \times 177}{283} = \frac{760 \times V_2}{273}$$

$$\therefore V_2 = \frac{759 \times 177 \times 273}{282 \times 760} = 170.5 \text{ ml}$$

Wt. of 170.5 ml of $H_2 = \frac{2}{22400} \times 170.5 \text{ g} = 0.0152 \text{ g}$

Now 0.0152 g H_2 is evolved from metal $= 0.444$ g

$$1 \text{ g } H_2 \text{ is evolved from metal} = \frac{0.444}{0.0152} = 29.21.$$

STEP II

Find the approximate Atomic Weight.

$$\text{Approx. At. wt.} \quad \frac{6.4}{\text{Sp. heat}} = \frac{6.4}{0.107} = 59.81$$

STEP III

Find the Valancy.

$$\text{At. wt.} = \text{Valency} \times \text{Eq. wt.}$$

or

$$\text{Valency} = \frac{\text{At. wt.}}{\text{Eq. wt.}} = \frac{59.81}{29.21} = 2.04$$

$$= 2 \text{ (whole number)}$$

STEP IV

Find the exact atomic weight.

$$\text{Exact At. wt.} = \text{Eq. wt.} \times \text{Whole no. Valency}$$
$$= 29.21 \times 2$$
$$= 58.42.$$

Vapour Density of Chloride Method. The atomic weight of an element is also found when the vapour density of the chloride of the element is given, in addition to its equivalent weight. This can be illustrated by the examples given below.

Atomic Weights

Example 6. *One gram of the chloride of an element was found to contain 0.835 g of chlorine. The vapour density of the chloride was 85. Find the atomic weight and molecular formula of the chloride.*

STEP I

Find the equivalent weight of the element.

$$\text{Wt. of chloride} = 1\cdot 0 \text{ g}$$
$$\text{Wt. of chlorine} = 0\cdot 835 \text{ g}$$

∴ Wt. of the element = $1\cdot 0 - 0\cdot 835 = 0\cdot 165$ g

Now, 0·835 g chlorine combines with metal = 0·165 g

∴ 35·5 g chlorine combines with metal

$$= \frac{0\cdot 165}{0\cdot 835} \times 35\cdot 5 = 7\cdot 015 \text{ g}$$

Thus eq. wt. = 7·015.

STEP II

Find the molecular weight of the chloride.

$$\text{Mol. wt.} = 2 \times \text{V. D.}$$
$$= 2 \times 85 = 170.$$

STEP III

Find the theoretical molcular weight of the chloride.

Let the valency of the element = x

∴ The formula of the chloride = ACl_x

where 'A' is some element.

Since atomic wt. = valency × eq. wt.

∴ Atomic wt. of the element = $x \times 7\cdot 015 = 7\cdot 015x$

∴ Molecular wt. of the chloride ACl_x
$$= 7\cdot 015x + 35\cdot 5x$$
$$= 42\cdot 51x$$

STEP IV

Find the valency.

The actual mol. wt. of the chloride = 170.

The theoretical mol. wt. = $42\cdot 51x$.

$$42.51x = 170$$

or $$x = \frac{170}{42.51} = 4.$$

STEP V

Find the atomic weight.

At. wt. = valency × eq. wt.

= 4 × 7·015 = 28·060

Thus, the atomic weight = 28·060

and the formula of the chloride = ACl_4.

Example 7. *The oxide of an element contains 53·3% oxygen. The V.D. of the chloride of the element is 85. Find the atomic weight of the element.*

STEP I

Find the equivalent weight of the element.

Wt. of oxide = 100 g
Wt. of oxygen = 53·3 g

∴ Wt. of the element = 100 − 53·3 = 46·7 g

Now 53·3 g oxygen combines with metal = 46·7 g

1 ,,   ,,   ,,   ,,   ,, = $\frac{46·7}{53·3}$ g

8 ,,   ,,   ,,   ,,   ,, = $\frac{46·7}{53·3} \times 8$ g

Thus equivalent wt. is = 7·007.

STEP II

Find the molecular weight of the chloride.

Mol. wt. = 2 × V.D. = 2 × 85 = 170.

STEP III

Find the molecular weight of the chloride.

Let the element be A.

∴ its chloride = ACl_x

where x is the valency of A.

At. wt. of the element A
$$= \text{Valency} \times \text{eq. wt.}$$
$$= x \times 7.007 = 7.007x$$

∴ theoretical mol. wt. of the chloride ACl_x
$$= 7.007x + 35.5x$$
$$= 42.507x$$

STEP IV

Find the valency.

Actual mol. wt. $= 170$
Theoretical mol. wt $= 42.507x$

∴ $42.507x = 170$

$$x = \frac{170}{42.507}$$
$$= 4.$$

STEP V

Find the atomic weight of the element.

At. wt. $= \text{Valency} \times \text{eq. wt}$
$$= 4 \times 7.007$$
$$= 28.028$$

Example 8. *The oxide of a metal contains 15% oxygen. 10 ml of the chloride of this metal diffuses in the same time as 16 ml of sulphur dioxide. Calculate the atomic weight of the metal.* (*Punjab Pre-Univ., 1981*)

STEP I

Calculate the Eq. wt. of the metal.

wt. of the oxide $= 100$ g
wt. of the metal $= 85$ g
wt. of oxygen $= 15$ g

Now 15 g of oxygen combines with 85 g of the metal

∴ 1 g ,, ,, will combine with metal $= \dfrac{85}{15}$ g

∴ 8 g ,, ,, ,, ,, ,, ,, $= \dfrac{85}{15} \times 8$ g
$\qquad\qquad\qquad\qquad\qquad\qquad = 45.3$ g

Hence eq. wt. of the metal = 45·3

STEP II

Calculate the mol. wt. of the metal chloride

Rate of diffusion of metal chloride $(r_1) = \dfrac{10 \text{ ml}}{t_1}$

where t_1 is the time of diffusion

rate of diffusion of sulphur dioxide, SO_2, $= \dfrac{16 \text{ m}}{t_2}$

where t_2 is the time of diffusion of SO_2.

but t_1 and t_2 are same, come say

$$= t.$$

$\therefore \quad r_1 = \dfrac{10}{t}$ and $r_2 = \dfrac{16}{t}$

Now $\quad \dfrac{r_1}{r_2} = \sqrt{\dfrac{M_2}{M_1}}$

$\therefore \quad \dfrac{10}{16} = \sqrt{\dfrac{\text{Mol. Wt. of } SO_2}{\text{Mol. Wt. of metal chloride}}}$

$$= \sqrt{\dfrac{64}{M_1 (M_{MCl})}}$$

(MCl is metal chloride)

Squaring both sides we have

$$\dfrac{10 \times 10}{16 \times 16} = \dfrac{64}{M_{MCl}}$$

$$M_{MCl} = \dfrac{64 \times 16 \times 16}{10 \times 10} = \dfrac{64 \times 64}{25}$$

$$= 163 \cdot 5$$

STEP III

Calculate the formula of the oxide

Let M be the metal whose valency $= x$

$\therefore$ Formula of metal chloride $= MCl_x$

STEP IV

Calculate the mol. wt. of MCl_x

From the atomic weight of elements

$$\text{Mol. wt. of } MCl_x = \text{At. wt. of } M + (x \times 35.5)$$
$$= (\text{Eq. wt.} \times \text{valency}) + 35.5\, x$$
$$= 45.3\, x + 35.5\, x$$
$$= 80.8\, x.$$

STEP V

Calculate the valency of the metal.

Equating molecular weights found in steps II and IV we get

$$80.5\, x = 163.5$$
$$\therefore x = \frac{163.5}{80.8}$$
$$= 2$$

STEP VI

Calculate the at. wt. of the Metal

$$\text{At. wt.} = \text{Eq. wt} \times \text{valency}$$
$$= 45.3 \times 2 = 90.6$$

END-OF-CHAPTER PROBLEMS

1. 0·2 g of divalent metal gave on treatment with a dilute mineral acid 68·4 ml of hydrogen at S.T.P. Calculate the atomic weight of the metal. [**Ans.** 65·50]

2. 1 g of a metal on treatment with nitric acid and subsequent ignition gave 1·27 g of its oxide. Find the atomic weight of the metal if its valency is 2. [**Ans.** 59·26]

3. 1·0 g of a monovalent metal yielded 1·328 g of its chloride. Calculate its atomic weight. [**Ans.** 108·2]

4. 1 g of the chloride of a trivalent element contains 0·405 g chlorine. Determine the atomic weight of the element. [**Ans.** 156·4]

5. 1·62 g of a divalent metal on ignition gave 2·02 g of its oxide. If the atomic weight of the metal be 65, find its eq. wt.
[**Ans.** 32.5]

Dulong and Petit's Method

6. The chloride of a metal was found on analysis to contain 79·77% chlorine. The specific heat of the metal was 0·237. Calculate the exact atomic weight of the metal. [**Ans.** 27·003]

7. 0·133 g of metal on treatment with dilute hydrochloric acid evolved 152 ml of moist hydrogen at 17°C and 740 mm pressure. The

specific heat of the metal is 0·25. Calculate the equivalent weight, atomic weight and valency of the metal. (Aq. tension at 17°C is 15 mm).
[**Ans.** 12·14, 24·28, 2]

8. The chloride of a metal contains 36·10% of the metal. Calculate the exact atomic weight if its specific heat is 0·16.
[**Ans.** 40·10]

9. An element forms an oxide containing 47% oxygen. The specific heat of the element is 0·22. Calculate the atomic weight, equivalent weight and valency of the element. Write the formula of the chloride, oxide and sulphate of the element.
[**Ans.** 27·066, 9·022, 3, MCl_3, M_2O_3, $M_2(SO_4)_3$]

10. The oxide of a metal contains 40% of oxygen and the valency of the metal is 2. Find the atomic weight of the metal.
[**Ans.** 24]

11. 0·45 g of a metal gave 176·6 cc of hydrogen at 23°C and 743 mm pressure when treated with dilute sulphuric acid. If the specific heat of the metal is 0·091, what would be the valency and exact atomic weight of the metal? (Aq. tension at 23°C is 21 mm).
[**Ans.** 2, 65·88]

12. 1·0 g of a metallic carbonate on heating gave 0·56 g of the metallic oxide. The specific heat of metal was 0·151. Calculate the eq. wt., valency and the correct atomic weight of the metal. Also give the formula of the carbonate taking 'M' as the symbol of the metal Given : equivalent weight of the carbonate radical =
[**Ans.** MCO_3, 20, 2, 40]

13. The chloride of a metal M contains 20·2% of the metal. Its specific heat is 0·224. What is the accurate atomic weight of the metal? If the vapour density of the chloride is 65·7, what is it its empirical formula?
[**Ans.** 26·95, MCl_5]

14. A metallic chloride contains 29% of the metal. If the specific heat of the metal is 0·15, find its atomic weight. (*Punjab H.S. 1970*)
[**Ans.** 43·5]

15. A metallic chloride contains 64% of chlorine. If the specific heat of the metal is 0·16, find its atomic weight. (*Punjab H.S. 1973*)
[**Ans.** 39·94]

16. 0·27 gm of a metal combines with 0·24 gm of oxygen to form an oxide. The sp. heat of the metal is 0·24. Calculate the at. wt. of the metal. (*Bihar Pre-Univ 1973*)
[**Ans.** 27]

17. A metallic oxide contains 28·5% oxygen. The sp. heat of the metal is 0·16. Find the equivalent weight and exact atomic weight of the metal. (*Shivaji Pre-Univ. 1975*)
[**Ans.** 20·07, 40·14]

18. 0·148 gm of a metal on treatment with dilute hydrochloric acid evolved 152 ml. of moist hydrogen at 27°C and 758 mm pressure. The sp. heat of the metal is 0·25. Calculate the equivalent weight,

the valency and the atomic weight of the metal. (Aq. T. at $27°C = 18$ mm) *(Guru Nanak Dev Pre-Univ., 1975)*
[**Ans.** 12·31, 2, 24·62]

19. The chloride of a metal M contains 20·2% of the metal. Its specific heat is 0·224. Calculate equivalent weight, valency, and exact Atomic weight of the metal.
(Punjabi Pre-Univ., 1978)
[**Ans.** Eq. wt. = 8·98
Valency = 3
At. wt. = 26·95]

20. A metal weighs 1·06 g and its oxide 1·33 g. The specific heat of the metal is 0·095: Calculate the exact atomic weight and valency of the metal. *(Punjab Board Hr. Sec., 1978)*
[**Ans.** At. wt. = 62·80
Valency = 2]

21. The oxide of an element contains 20% of oxygen. The specific heat of the element is 0·068. Calculate the equivalent weight and atomic weight of the element. *(Haryana Board Hr. Sec. 1979)*
[**Ans.** Eq. wt. = 32
At. wt = 96·00]

22. 1·0 g of a metallic carbonate left on ignition 0·56 g of its oxide. What is the atomic weight of the metal if its specific heat is 0·155. *(Guru Nanak Dev Pre-Univ., 1981)*
[**Ans.** Exact atomic weight=40]

23. The oxide of a metal contains 15% oxygen. 10 ml of the chloride of this metal diffuse in the same time as 16 ml of sulphur dioxide. Calculate the atomic weight of the metal.
(Punjab Pre-Univ., 1981)
[**Ans.** 90·6]

24. 1·05 g of a metal combine with 0·70 g of oxygen. Calculate the atomic weight of the metal if it is divalent.
(Punjabi Pre-Univ., 1981)
[**Ans.** 24]

25. The chloride of an element contains 36·1% of the metal. Calculate the exact atomic weight of the metal if the specific heat is 0·16. *(Himachal Pre-Univ., 1981)*
[**Ans.** 40]

26. 0·20 g of a tri-valent metal combines with exactly 50 ml of oxygen measured at NTP to form its oxide. Calculate the atomic weight of the metal. *(Kurukshetra Pre-Univ., 1981)*
[**Ans.** 144·4]

From Vapour Density of the Chloride of the Element

27. The chloride of a metal M is found to contain 20·2 per cent of the metal. The specific heat of the metal is 0·224. What is the accurate atomic weight ? If the V.D. of the chloride is 66·7, what is the molecular formula of the chloride ?

[Ans. 26·96, MCl_4]

28. A metal on being heated in a current of chlorine yields a liquid chloride containing 54·4% of chlorine. 0·522 g of this compound gave 47·6 cc of vapours at 15°C and 754 mm pressure. Calculate the atomic weight of the metal and find the molecular formula of the chloride.
(Ranchi Pre-Univ., 1974)

[Ans. 119·0 MCl_4]

29. The oxide of a metal was found to contain 47·06% of oxygen. If the vapour density of chloride of the metal is 66·75. Calculate the atomic weight of the metal.
(Punjab Pre-Univ., 1974)

[Ans. 26·625]

30. 0·148 g of a metal on treatment with dilute hydrochloric acid evolved 152 ml of moist hydrogen at 27°C and 758 mm pressure. The specific heat of the metal is 0·25. Calculate the equivalent weight, valency and the atomic weight of the metal. (Aqueous tension at 27°C = 18 mm).
(Guru Nanak Dev Pre-Univ., 1975)

[Ans. 12·39, 2. 24·78]

31. The equivalent weight of a metal (M) of specific heat 0·23 is 8·99. Write down the formulae of its chloride, sulphate and nitrate.
(Madurai Pre-Univ., 1975)

[Ans. MCl_3, $M_2(SO_4)_3$, $M(NO_3)_3$]

32. The specific heat of a metal M is 0·25 and its equivalent weight is 12. What is its correct atomic weight ? Give the formula of its chloride.
(Gauhati Pre-Univ., 1975)

[Ans. 24; MCl_2]

33. The chloride of a trivalent metal contains 45·6 per cent chlorine. Find atomic weight of the metal.
(Kashmir Pre-Univ 1975)

[Ans. 127·1]

34. A metallic chloride contains 34·4 per cent of the metal. The specific heat of the metal is 0·11. Calculate the exact atomic weight of the metal.
(Punjab Univ Pre-Univ., 1976)

[Ans. 55·86]

Miscellaneous

35. The equivalent weight of an element is 5·0. It forms a chloride whose vapour density is 60. Calculate the valency and the atomic weight.
(Punjabi Pre-Univ., 1980)

[Ans. valency = 3; atomic weight = 15]

36. 0·45 g of metal on treatment with excess of dilute acid gave 176·6 ml of moist hydrogen measured at 23°C and 743·0 mm pressure.

The specific heat of the metal is 0·091. Calculate exact atomic weight of the metal. (Aqueous tension at 23°C = 21 mm).

(*Punjabi Pre-Univ., 1980*)

[**Ans.** Exact atomic weight = 65·14]

37. Calculate the mass of an electron in atomic mass unit. (mass of electron = 9.11×10^{-28} g). (*Punjabi Pre-Univ., 1980*)

[**Ans.** 5.487×10^{-5}]

38. The chloride of a metal M contains 20·2% of the metal. The specific heat of the metal is 0·224. Calculate the accurate atomic weight of M.

If the V.D. of the above chloride is 66·7, what is its molecular formula ? (*Calicut Pre-degree, 1980*)

[**Ans.** atomic weight = 26·94
molecular formula = MCl_3]

39. The oxide of a metal contains 15% oxygen. 10 ml of the chloride of this metal diffuse in the same time as 16 ml of sulphur dioxide. Calculate the atomic weight of the metal.

(*Panjab Pre-Univ. 1981*)

[**Ans.** 90·6]

11

Oxidation and Reduction

Reactions in which electrons are transferred are called *oxidation-reduction* or **redox reactions**. *Oxidation* is an apparent loss of electrons by an atom, ion or molecule while *reduction* is an apparent gain of electrons. Oxidation and reduction always proceed simultaneously and the total number of electrons lost in oxidation must equal the number of electrons gained in the reduction.

Example 1. *Reduction of a ferrous salt with chlorine resulting into the formation of ferric chloride involves both oxidation and reduction. Ferrous is oxidised to ferric while chlorine is reduced to chloride ion. The overall reaction is*

$$2FeCl_2 + Cl_2 \longrightarrow 2FeCl_3$$
Ferrous chloride $\qquad$ Ferric chloride

This overall reaction may be split into two reactions

(i) $\qquad 2Fe^{2+} - 2e \longrightarrow 2Fe^{3+}$ —this is oxidation

(ii) $\qquad Cl_2 + 2e \longrightarrow 2Cl^-$ —this is reduction

2. *Oxidation of magnesium with oxygen*

$$2Mg + O_2 \longrightarrow 2MgO$$

This reaction may be split into two reactions

(i) $\qquad Mg - 2e \longrightarrow Mg^{2+}$ —this is oxidation

(ii) $\qquad O_2 + 4e \longrightarrow 2O^{2-}$ —this is reduction

Oxidation and Reduction

This may also be represented as

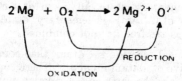

3. Formation of sodium chloride

$$Na(s) + \tfrac{1}{2}Cl_2(g) \longrightarrow NaCl$$

Sodium is oxidised and chlorine is reduced. This equation may be split as

(i) $\quad Na - e \longrightarrow Na^+$ —this is oxidation
(ii) $\quad Cl + e \longrightarrow Cl^-$ —this is reduction

Oxidation Number. We often assign positive and negative numbers to some of the elements e.g., hydrogen has valency number $+1$ and oxygen -2. The negative charge means excess of electrons and positive charge results when there is a deficiency of electrons. In order to determine the number of electrons lost or gained in a chemical reaction, an integer called oxidation number is assigned to each atom, in a molecule or ion. This number expresses the *oxidation state* of the atom concerned. The change in oxidation state of an atom determines whether or not a reaction involves oxidation or reduction. Thus in a reaction between chlorine and water, the oxidation number of various elements involved are designated as superscripts.

$$\overset{0}{Cl_2} + \overset{+1}{H_2}\overset{-2}{O} \longrightarrow \overset{+}{H}\,\overset{-1}{Cl} + \overset{+}{H}\,\overset{+}{Cl}\,\overset{-2}{O}$$

Similarly, in the reaction between copper and dil. nitric acid we have

$$2\overset{0}{Cu} + 8\overset{+1}{H}\,\overset{+5}{N}\,\overset{-2}{O_3} \longrightarrow 3\overset{+2}{Cu}\,(\overset{+5\,-2}{NO_3})_2 + 2\overset{+2}{N}\,\overset{-2}{O}$$

$$+ 4\overset{+1}{H_2}\,\overset{-2}{O}$$

In the reaction between $AgNO_3$ and $NaCl$ we have

$$\overset{+1}{Ag}\,\overset{+5}{N}\,\overset{-2}{O_3} + \overset{+1}{Na}\,\overset{-1}{Cl} \longrightarrow \overset{+1}{Ag}\,\overset{-1}{Cl} + \overset{+1}{Na}\,\overset{+5}{N}\,\overset{-2}{O_3}$$

Rules for finding oxidation numbers.

1. In the elementary state an atom or a molecule has an oxidation number of 0 (zero). Elementary state means when atoms or molecules are free *i.e.* not combined.

2. In a chemical reaction the atom that combines assumes positive or negative oxidation number. These numbers, for all pratical purposes, are the same as the valency numbers assigned to each element.

3. During the course of a chemical reaction, if an atom acquires a positive number, it is oxidised and if the change is a decrease in positive number, or if the negative number increases, it is reduction.

4. When hydrogen is present in a compound, it is assigned an oxidation number $+1$ (except in hydrides where it is -1).

5. Oxygen always assumes a -2 oxidation number except when it is in the form of peroxide O_2^{2-}, when it has oxidation number -1 (it has oxidation no. -1 even when it combines with fluorine).

6. Oxidation number of the metals is usually positive since they most often combine with more negative non-metals.

7. The sum of all oxidation numbers in an ion or a radical is equal to the charge on the ion or the radical.

8. In the course of an oxidation-reduction reaction, the increase of positive oxidation number by one atom must always be balanced by a decrease in number by another atom, so that the change in oxidation and reduction is equal

Let us now determine the oxidation number of elements in their compounds.

Example 1. *What is the oxidation number of each element in (i) H_2O ; (ii) $CaCl_2$; (iii) H_2O_2 ?*

(*i*) **H_2O.** Oxidation number of $H = +1$

Since there are two H atoms, therefore, total value of the oxidation number of hydrogen $= 2 \times (+1) = +2$.

Let the oxidation number of oxygen $= x$

Oxidation and Reduction

Since the sum of the oxidation nos. of hydrogen and oxygen in a compound must be zero.

$$\therefore \quad +2+x=0$$

or $\quad x=-2.$

Thus, oxidation number of oxygen $=-2$.

(ii) $CaCl_2$

Oxidation number of chlorine when it combines with a metal $=-1$.

There are two Cl atoms

$\therefore$ total oxid. no. of $Cl = 2 \times (-1) = -2$

Let the oxid. no. of $Ca = x$

Since the sum must be zero i.e. $x - 2 = 0$

or $\quad x = +2$

Hence, **oxidation number of calcium is $+2$.**

(iii) H_2O_2

Oxidation no. of hydrogen $= +1$

For two atoms, the value $= 2 \times (+1) = +2$.

Let the oxidation no. of oxygen $= x$

Since $\quad x + 2 = 0$

$\therefore \quad x = -2.$

Hence the oxidation number of oxygen in $H_2O_2 = -1$.

Example 2. *Find the oxidation number of chlorine in $KClO_3$. Given the oxidation number of $K = +1$ and $O = -2$.*

Let the oxidation number of $Cl = x$

Since the sum of oxidation numbers in a compound must be zero.

i.e., $\quad K + Cl + 3O = 0$

or $\quad +1 + x + 3 \times (-2) = 0$

or $\qquad +1+x-6=0$

∴ $\qquad x=-5$

Hence the oxidation no. of chlorine in $KClO_3 = -5$.

Example 3. *Find the oxidation no. of Sn in the compound Na_2SnO_2. Given oxidation number of $Na = +1$, $O = -2$.*

Let the oxidation no. of $Sn = x$

Since the algebraic sum of oxidation nos. in a compound is zero, we have.

$$2Na + Sn + 2O = 0$$

∴ $\qquad 2 \times (+1) + x + 2 \times (-2) = 0$

or $\qquad +2 + x - 4 = 0$

or $\qquad x - 2 = 0$

or $\qquad x = +2$

∴ oxidation number of Sn in $Na_2SnO_2 = +2$

Balancing of oxidation reaction by oxidation state method. We can balance oxidation-reduction reactions by oxidation state method as per rules given below ·

1. First write the skeleton equation.

2. Write the oxidation number of each element in the reactants and products as superscripts.

3. Pick out the molecule, ion or radical containing the element that undergoes oxidation and the molecule, ion or radical containing the same element in its oxidised form. This pair forms the basis of the oxidation reaction.

4. Select a similar pair that undergoes reduction. This forms the basis of the reduction reaction.

5. Balance the individual oxidation and reduction reactions first chemically and then with respect to electrons.

Example 1. *Balance the equation.*

$$Cu + HNO_3 \longrightarrow Cu(NO_3)_2 + NO + H_2O.$$

The skeleton equation is already given. Write oxidation numbers of elements as superscripts.

Oxidation and Reduction

$$\overset{0}{Cu} + \overset{+1\,+5\,-2}{H\,N\,O_3} \longrightarrow \overset{+2\,\,+5\,\,-2}{Cu(NO_3)_2} + \overset{+2\,-2}{NO} + \overset{+1\,-2}{H_2O}$$

Copper changes from 0 to $+2$ (or it loses electrons), so it is being oxidised. Thus,

$$Cu - 2e \longrightarrow Cu^{+2} \quad \text{oxidation reaction}$$

Nitrogen changes from $+5$ to $+2$ (or it gains electrons), so it is being reduced. Thus,

$$\overset{+5}{N} + 3e \longrightarrow \overset{+2}{N}$$

Thus the two oxidation-reduction reactions are:

$$\overset{0}{Cu} - 2e \longrightarrow \overset{+2}{Cu} \qquad \text{oxidation} \qquad \ldots(1)$$

$$\overset{+5}{N} + 3e \longrightarrow \overset{+2}{N} \qquad \text{reduction} \qquad \ldots(2)$$

Multiply (1) by 3 and (2) by 2 we have

$$3Cu - 6e \longrightarrow 3Cu^{+2}$$

$$2N + 6e \longrightarrow 2N$$

The loss and gain of electrons in the two equations is thus balanced. On this basis now multiply copper by **3** and nitric acid by **2** in the skeleton equation. Then,

$$3Cu + 2HNO_3 \longrightarrow 3Cu(NO_3)_2 + 2NO + H_2O$$

Now we find that for 1 Cu atom 2 nitric acid molecules are required to form its nitrate, therefore, for 3Cu atoms we shall require 6 nitric acid molecules. 2 nitric acid molecules we already need for balancing the loss and gain of electrons, so total 8 nitric acid molecules will be needed. Thus,

$$3Cu + 8HNO_3 \longrightarrow 3Cu(NO_3)_2 + 2NO + H_2O$$

There are 8H atoms in nitric acid, these will produce $4H_2O$ molecules. Hence the balanced equation is

$$3Cu + 8HNO_3 \longrightarrow 3Cu(NO_3)_2 + 2NO + 4H_2O$$

Example 2. *Balance the equation*

$$KMnO_4 + KCl + H_2SO_4 \longrightarrow K_2SO_4 + MnSO_4 + H_2O + Cl_2$$

The skeleton equation is already given. Here $KMnO_4$ is the oxidising agent because Mn undergoes a change in oxidation from $+7$ to $+2$ as is seen from the oxidation numbers.

$$\overset{+1+7-2}{KMnO_4} + \overset{+1\ -1}{KCl} + \overset{+1+6-2}{H_2SO_4} \longrightarrow$$

$$\overset{+1+6-2}{K_2SO_4} + \overset{+2+6-2}{MnSO_4} + \overset{+1-2}{H_2O} + \overset{0}{Cl_2}$$

The oxidation state of Cl changes from -1 to 0. This is increase of oxidation number and hence chloride ion is oxidised to chlorine.

The two reactions involved are :

(i) $\overset{+7}{Mn} + 5e \longrightarrow \overset{+2}{Mn}$ Reduction

(ii) $2\overset{-1}{C} - 2e \longrightarrow Cl_2$ Oxidation

To balance the gain and loss of electrons, multiply equation (i) by 2 and (ii) by 5. Thus :

$$\overset{+7}{2Mn} + 10e \longrightarrow \overset{+2}{2Mn}$$

$$10Cl - 10e \longrightarrow 5Cl_2$$

Now multiply $KMnO_4$ (which contains Mn) by 2 and KCl (which contains chloride ion) by 10. We get

$$2KMnO_4 + 10KCl + H_2SO_4 \longrightarrow 2MnSO_4 + K_2SO_4 + H_2O + Cl_2$$

1 Mn atom forms 1 $MnSO_4$, so we need $2H_2SO_4$ molecules to produce $2MnSO_4$ from 2Mn. Since 2 K atoms require $1H_2SO_4$ to form $1K_2SO_4$ molecule, so 12 K atoms would require $6H_2SO_4$. Hence total H_2SO_4 molecules required are 8. Thus we may write.

$$2KMnO_4 + 10KCl + 8H_2SO_4 \longrightarrow 2MnSO_4 + 6K_2SO_4 + 8H_2O + 5Cl_2$$

Since there are 10 Cl in 10KCl they will produce $5Cl_2$ hence multiply Cl_2 on R.H.S. by 5. Similarly there are $8H_2SO_4$ molecules, they will produce $8H_2O$ molecules. Thus, the balanced equation is

$$2KMnO_4 + 10KCl + 8H_2SO_4 \longrightarrow 2MnSO_4 + 6H_2SO + 8H_2O + 5Cl_2$$

Example 3. *Balance the equation*

$$H_2S + KMnO_4 + H_2SO_4 \longrightarrow KHSO_4 + MnSO_4 + S + H_2O$$

The skeleton equation is already given. Write the oxidation number of elements.

$$\overset{+1-2}{H_2S} + \overset{+1+7-2}{KMnO_4} + \overset{+1+6-2}{H_2SO_4} \longrightarrow$$

$$\overset{+1+1+6-2}{KHSO_4} + \overset{+2+6-2}{MnSO_4} + \overset{0}{S} + \overset{+1-2}{H_2O}$$

Here sulphur changes from -2 to 0 hence oxidised and Mn changes from $+7$ to $+2$, hence reduced. We write these changes as

$$\overset{+7}{Mn} + 5e \longrightarrow \overset{+2}{Mn} \qquad \ldots(1)$$

$$\overset{-2}{S} - 2e \longrightarrow \overset{0}{S} \qquad \ldots(2)$$

To balance the gain and loss of electrons, multiply equation (1) by 2 and (2) by 5. We have,

$$\overset{+7}{2Mn} + 10e \longrightarrow \overset{+2}{2Mn}$$

$$\overset{-2}{5S} - 10e \longrightarrow 5S$$

Hence coefficients of $KMnO_4$ and $MnSO_4$ are 2, that of H_2 is 5. We now write,

$$2KMnO_4 + 5H_2S + H_2SO_4 \longrightarrow 2KHSO_4 + 2MnSO_4 + 5S + H_2O.$$

For 1 Mn we require $1H_2SO_4$ molecule to convert it to $MnSO_4$ and for 2K atom 2 more H_2SO_4 are required to convert these to $2KHSO_4$. Thus total H_2SO_4 required are 4.

$$2KMnO_4 + 5H_2S + 4H_2SO_4 \longrightarrow 2KHSO_4 + 2MnSO_4 + 5S + H_2O$$

There are 8H in $4H_2SO_4$ molecules. These will change to $4H_2O$'s. Hence the final equation is

$$2KMnO_4 + 5H_2S + 4H_2SO_4 \longrightarrow 2KHSO_4 + 2MnSO_4 + 5S + 4H_2O.$$

Example 4. *Write a balanced equation for the reaction of $\overset{+2}{Fe}$ with MnO_4^-. The products are $\overset{3+}{Fe}$ and $\overset{2+}{Mn}$.*

The equations involved are

(i) $\overset{2+}{Fe} - e \longrightarrow \overset{3+}{Fe}$ Oxidation

(ii) $\overset{+7}{Mn}\overset{2}{O_4} + 5e \longrightarrow \overset{+2}{Mn}$ Reduction

$4O^{-2}$ in MnO_4^- required $8H^+$ to form H_2O. Also to make up the loss and gain of electrons, multiply (i) by 5. Thus the coefficient of Fe^{2+} is 5. Thus

$$5Fe^{2+} - 5e \longrightarrow 5Fe^{3+}$$
$$8H^+ + MnO_4^- + 5e \longrightarrow Mn^{2+} + 4H_2O$$
$$\overline{5Fe^{2+} + 8H^+ + MnO_4^- \longrightarrow 5Fe^{3+} + Mn^{2+} + 4H_2O}$$

Example 5. *Give the balanced equation for the oxidation of oxalic acid by potassium permanganate in acid solution.*

Skeleton equation

$$H^+ + \overset{+7\ -2}{MnO_4^-} + \overset{+1\ +3\ -2}{H_2C_2O_4} \longrightarrow \overset{+4\ -2}{CO_2} + \overset{+2}{Mn}$$

Oxidation $\overset{+3}{H_2C_2O_4} - 2e \longrightarrow 2\overset{+4}{CO_2}$ ...(1)

Reduction $\overset{+7}{MnO_4^-} + 5e \longrightarrow \overset{+2}{Mn}$ ...(2)

Multiply (1) by 5 and (2) by 2 we have the coefficients for oxalic acid 5 and for MnO_4^- as 2. Thus

$$5H_2C_2O_4 + 2MnO_4 \longrightarrow 10CO_2 + 2Mn^{2+}$$

8 oxygen atoms in $2MnO_4^-$ ions will require $16H^+$ to form $8H_2O$. But $10H$ are already present in $5H_2C_2O_4$ molecules, so we require only $6H^+$ more. Hence the balanced equation is :

$$6H^+ + 2MnO_4^- + 5H_2C_2O_4 \longrightarrow 10CO_2 + 2Mn^{2+} + 8H_2O.$$

END-OF-CHAPTER PROBLEMS

1. Find the oxidation number of elements in the following compounds ;

(i) SO_2 ; (ii) NO_2 ; (iii) HSO_3^- ; (iv) BF_4^- ;
(v) Li_2SO_4 ; (vi) K_2O_2 ; (vii) Se^{2+} ; (viii) AsO_4^{3-}

Oxidation and Reduction 147

(ix) $AsSO_3^{-3-}$; (x) K_2SO_4.

[**Ans.** (i) $\overset{+4-2}{SO_2}$ (ii) $\overset{+4-2}{NO_2}$ (iii) $\overset{+1+4-2}{HSO_3}$

(iv) $\overset{+3-1}{BF_4}$ (v) $\overset{+1+6-2}{Li_2SO_4}$ (vi) $\overset{+2-1}{K_2O_2}$

(vii) $\overset{+3}{Se}$ (viii) $\overset{+5-2}{AsO_4}$ (ix) $\overset{+3-2}{AsO_3}$

(x) $\overset{+1\ +6\ -3}{K_2\ S\ O_4}$]

2. What is the oxidation number of chromium in

(i) Cr_2O_3 (ii) $K_2Cr_2O_7$ (iii) $CrCl_3$ (iv) $CrSO_4$
(v) $Cr(OH)_3$ (iv) CrO.

[**Ans.** (i) +3 (ii) +6 (iii) +3 (iv) +2
(v) +3 (vi) +2].

3. What is the oxidation number of manganese in

(i) K_2MnO_4 (ii) $KMnO_4$ (iii) MnO_4^- (iv) Mn_2O_7
[**Ans.** (i) +6 (ii) +7 (iii) +7 (iv) +7]

4. What is the oxidation number of nitrogen in

(i) HNO_2 (ii) HNO_3 (iii) N_2 (iv) N_2O (v) NO
(vi) N_2O_5 (vii) NH_3.
[**Ans.** (i) +3 (ii) +5 (iii) 0 (iv) +1
(v) +2 (vi) +5 (vii) +3]

5. Balance the following equations by oxidation state method.

(i) $FeCrO_4 + K_2CO_3 + O_2 \longrightarrow Fe_2O_3 + K_2CrO_4 + CO_2$

(ii) $C_{12}H_{22}O_{11} + H_2SO_4 \longrightarrow CO_2 + SO_2 + H_2O$

(iii) $NaBr + KMnO_4 + H_2SO_4 \longrightarrow Br_2 + MnSO_4 + Na_2SO_4 + K_2SO_4 + 8H_2O$

(iv) $SO_2 + HNO_3 + H_2O \longrightarrow H_2SO_4 + NO$

(v) $Al + K_2Cr_2O_7 + H_2SO_4 \longrightarrow Al_2(SO_4)_3 + Cr_2(SO_4)_3 + K_2SO_4 + H_2O$

(vi) $KMnO_4 + K_2SO_3 + H_2O \longrightarrow MnO_2 + K_2SO_4 + KOH$

(vii) $H_2SO_3 + H_2S \longrightarrow S + H_2O$

(viii) $K_2Cr_2O_7 + H_2O + S \longrightarrow SO_2 + KOH + Cr_2O_3$

(ix) $KClO_3 \longrightarrow KCl + O_2$

(x) $MnO_2 + HCl \longrightarrow MnCl_2 + Cl_2 + H_2O$.

[**Ans.** (i) $4FeCr_2O_4 + 8K_2CO_3 + 7O_2 \longrightarrow 2Fe_2O_3 + 8K_2CrO_4 + 8CO_2$

(ii) $C_{12}H_{22}O_{11} + 24H_2SO_4 \longrightarrow 12CO_2 + 24SO_2 + 35H_2O$

(iii) $10NaBr + 2KMnO_4 + 8H_2SO_4 \longrightarrow 5Br_2 + 2MnSO_4 + 5Na_2SO_4 + K_2SO_4 + 8H_2O$

(iv) $3SO_2 + 2HNO_3 + 2H_2O \longrightarrow 3H_2SO_4 + 2NO$

(v) $2Al + K_2Cr_2O_7 + H_2SO_4 \longrightarrow$
$\quad\quad Al_2(SO_4)_3 + Cr_2(SO_4)_3 + K_2SO_4 + 7H_2O$

(vi) $2KMnO_4 + 3K_2SO_3 + H_2O \longrightarrow$
$\quad\quad 2MnO_2 + 3K_2SO_4 + 2KOH$

(vii) $H_2SO_3 + 2H_2S \longrightarrow 3S + 3H_2O$

(viii) $2K_2Cr_2O_7 + H_2O + 3S \longrightarrow 3SO_2 + KOH + 2Cr_2O_3$

(ix) $2KClO_3 \rightarrow 2KCl + 3O_2$

(x) $MnO_2 + 4HCl \longrightarrow MnCl_2 + Cl_2 + 2H_2O$].

6. Balance the half reaction for the reduction of $H_3AsO_4 \longrightarrow HAsO_3$ in acidic solution.

[**Ans.** $\overset{+1\;+5\;-2}{H_3AsO_4} \longrightarrow \overset{+1\;+3\;-2}{HAsO_2}$

$H_3AsO_4 + 2e \longrightarrow HAsO_2$

$2H^+ + 2e + H_3AsO_4 \longrightarrow$
$\quad\quad HAsO_2 + 2H_2O$]

7. Assign oxidation numbers in the reaction

$Cl_2 \longrightarrow Cl^- + HClO$

[**Ans.** $\overset{0}{Cl_2} \longrightarrow \overset{-1}{Cl} + \overset{+1}{H}\overset{+1}{Cl}\overset{-2}{O}$]

8. Balance the following equations by half reaction method :

(i) $Ni(s) + Mn_2O(s) + \overset{+}{H}(aq) \rightarrow \overset{2+}{Ni}(aq) + \overset{2+}{Mn}(aq) + H_2O(l)$

(ii) $Zn(s) + Cl_2(g) \rightarrow \overset{2+}{Zn}(aq) + Cl(aq)$

(Punjabi Pre-Univ., 1973)

[**Ans.** (i) $Ni(s) + MnO_2(s) + 4H^+(aq) \rightarrow Ni^{2+}(aq) + Mn^{2+}(aq) + 2H_2O(l)$

(ii) $Zn(s) + Cl_2(g) = Zn^{2+}(aq) + 2Cl^-(aq)$]

9. By use of ion-electron (half reaction) method, give a balanced equation for each of the following reactions :

(i) $\overset{2-}{Cr_2O_7} + \overset{2+}{Fe} + H^+ \longrightarrow Cr^{3+} + \overset{3+}{Fe} + H_2O(l)$

(ii) $H_2O_2 + I^- + H^+ \longrightarrow H_2O + I_2$

(iii) $AsO_3^{3-} + IO_3^- \longrightarrow AsO_4^{3-} + I^-$ (in acidic medium)

(iv) $Cr_2O_7^{2-} + H^+ + Cl^- \longrightarrow Cr^{3+} + Cl_2 + H_2O$

(v) $MnO_4^- + Fe^{2+} + H^+ \longrightarrow Mn^{2+} + Fe^{3+} + H_2O$
(Kurukshetra Pre-Univ., 1974)

[**Ans.** (i) $Cr_2O_7^{2-} + 6Fe^{2+} + 14H^+ \longrightarrow 2Cr^{3+} + 6Fe^{3+} + 7H_2O$,)

(ii) $H_2O_2 + 2I^- + 2H^+ \longrightarrow 2H_2O + I_2$

(iii) $3AsO_3^{3-} + IO_3^- \longrightarrow 3AsO_4^{3-} + I^-$

(iv) $Cr_2O_7^{2-} + 14H^+ + 6Cl^- \longrightarrow 2Cr^{3+} + 3Cl_2 + 7H_2O$

(v) $MnO_4^- + 5Fe^{2+} + 8H^+ \longrightarrow Mn^{2+} + 5Fe^{3+} + 4H_2O$]

10. What is the oxidation number of

(i) S in $Na_2S_2O_3$ (ii) N in NO_3^- (iii) Mn in $KMnO_4$
(Punjab Pre-Univ., 1977)
[**Ans.** $+2$, $+5$, $+7$]

11. What is the oxidation number of N, S, Cl, and Cr in HNO_3; Na_2SO_4; $KClO_3$; and $K_2Cr_2O_7$ respectively.
(Himachal Pre-Univ., 1977)
[**Ans.** $+5, +6, +5,$ and $+6$]

12. What is the oxidation number of the following:

(a) Carbon in Na_2CO_3

(b) Cr in $Cr_2O_7^{-2}$

(c) Mn in MnO_4^{-1}

(d) Mn in MnO_4^{-2}

(e) S in Na_2SO_3

(Guru Nanak Dev Pre-Univ., 1978)
[**Ans.** (a) $+4$;   (b) $+6$;   (c) $+7$;
(d) $+6$;   (e) $+4$]

13. Calculate the oxidation number of:

(a) S in $Na_2S_2O_3$

(b) Mn in K_2MnO_4

(c) P in P_2O_5

(Himachal Pre-Univ., 1978)
[**Ans.** (a) $+2$;   (b) $+6$;   (c) $+5$]

14. Balance the following equations by oxidation state method:
(a) $NH_3 + O_2 \longrightarrow NO_2 + H_2O$

(b) $HNO_3 + I_2 \longrightarrow HIO_3 + NO_2 + H_2O$

(Himachal Pre-Univ., 1978)

[**Ans.** (a) $4NH_3 + 7O_2 \longrightarrow 4NO_2 + 6H_2O$

(b) $10 HNO_3 + I_2 \longrightarrow 2HIO_3 + 10NO_2 + 4H_2O$]

15. (a) Calculate the oxidation number of Nitrogen in N_2O_5 and carbon in CH_3Cl.

(b) Balance the following equation by Ion Electron (Half-Reaction) method :

$MnO_4^{-1} + Fe^{+2} + H^+ \longrightarrow Fe^{+3} + Mn^{+2} + H_2O$

(Haryana Board Hr. Sec., 1979)

[**Ans.** (a) $+5, +4$

(b) $MnO_4^{-1} + 5 F^{+2} + 8H^+ \longrightarrow Mn^{+2} + 5Fe^{+3} + 4HO_2$]

Volumetric Analysis

CONCENTRATION

A solution of one substance in another consists of *a Solute*, the dissolved substance and the *solvent* in which the solute has been dissolved. The *concentration* of a solution is the amount of, the solute dissolved per unit volume or weight of the solvent. The concentration of a solution can be expressed in a number of ways :

(a) Grams of the solute per 100 g of the solvent.
(b) Grams of the solute per 100 ml of the solvent.
(c) Moles of the solute per litre of the solvent.
(d) Moles of the solute per 1000 g of the solvent.

A standard solution. *A standard solution is one whose concentration is known.*

Normal solution. *A normal solution is one which is prepared by dissolving 1 gram equivalent of the solute per 1000 ml of the solution.* Thus one normal (1N) solution of NaOH contains 1 g equivalent (40 g) of NaOH dissolved per litre of the NaOH solution. Similarly one half normal (0·5N) or semi normal solution of HCl will contain 18·25 g of HCl dissolved per litre of HCl solution. $\frac{N}{10}$ KMnO$_4$ solution means the solution of KMnO$_4$ in which $\frac{1}{10}$ of the gram equivalent weight of KMnO₄ (3·16 g) has been dissolved per litre of the KMnO₄ solution. Thus **normality**. *of a solution is the number of gram-equivalent weights of the solute contained in one litre of the solution*. The normality of a solution is denoted by the letter N.

Thus normality of a solution

$$= \frac{\text{gram equivalents of the solute}}{\text{litres of the solution}}$$

Formal solution or Formality. *A formal solution is one which has been obtained by dissolving one formula weight of the solute in one litre of the solution.* The formality of solution is denoted by F. Thus 1 F solution of oxalic acid means which has been prepared by dissolving 126 g

$$\begin{array}{c}\text{COOH}\\|\qquad\quad 2\text{H}_2\text{O}\\\text{COOH}\end{array}$$

i.e., the number of grams equal to the formula weight of oxalic acid in one litre solution. Similiary 1F HCl solution means, the solution of HCl which contains the number of grams equal to the formula weight (*i.e.*, 36·5 g) of HCl dissolved in one litre solution. Thus *formality of a solution is the number of gram-formula weights of the solute contained in one litre solution*. When we say 1 molar solution or 1 Formal solution of NaOH it is the same thing. (*i.e.*, 40 g NaOH dissolved in 1 litre of solution). Molarity is expressed by the letter M.

Thus Molarity or Formality

$$= \frac{\text{No. of gram-formula wts. of the solute}}{\text{Volume in litres of the solution}}$$

The molarity or formality of a solution is always fixed since the formula weight or the molecular weight of a substance does not change with the reaction (while normality of a solution may change with the nature of the reaction).

Molal Solution. *A* **molal solution** *is one which is obtained by dissolving one gram molecule of the solute in* **1000 grams** *of the solvent.* The molality of a solution is expressed by the letter '*m*'. The molality of a solution is independent

Volumetric Analysis

of temperature and is hence useful for experiments involving measurements of depression in freezing point, boiling point (elevation), lowering of vapour pressure etc. (Normality, formality or molarity is defined in terms of volume of the solution which changes with the change in temperature).

Thus, molality of a solution $= \dfrac{\text{Moles of the solute}}{\text{kilogram of the solvent}}$

We may thus summarize :

$Normality\ (N) = \dfrac{\text{Gram-equivalents of the solute}}{\text{Volume in litres of the solution}}$

$Formality\ (F) = \dfrac{\text{No. of g-formula weight of the solute}}{\text{Volume of the solution in litres}}$

$Molarity\ (M) = \dfrac{\text{No. of moles of the solute}}{\text{Volume of the solution in litres}}$

$Molality\ (m) = \dfrac{\text{No. of moles of the solute}}{\text{Weight of the solvent in kilograms}}$

Example 1. *How many grams of KOH are required to prepare* $1N, 2N, \dfrac{N}{10}, \dfrac{N}{2}$ *solutions ?*

Step I

Find the equivalent weight of KOH.

The equivalent weight of KOH

$= \dfrac{\text{Mol. wt.}}{\text{No. of replaceable OH groups}}$

$= \dfrac{56}{1} = 56.$

Step II

To prepare 1N KOH, we need 1 g equivalent weight

i.e., $\quad 1 \times 56 = 56$ g for 1 litre solution.

To prepare 2N solution, we need 2 g equivalent weights

i.e., $\quad 2 \times 56 = 112$ g for 1 litre solution.

To prepare $\frac{N}{10}$ KOH solution we need $\frac{1}{10}$ g-equivalent weight of KOH

i.e., $\frac{1}{10} \times 56 = 5.6$ g for the one litre solution.

To prepare $\frac{N}{2}$ KOH solution, we need $\frac{1}{2}$ g-equivalent weight of KOH

i.e., $\frac{1}{2} \times 56 = 28$ g for one litre solution.

Example 2. *How many grams of H_2SO_4 are required to prepare 1N, 2N, 1F, 2F, 1M, 2M, $\frac{M}{2}$ solutions of H_2SO_4?*

STEP I

Find the equivalent weight of H_2SO_4.

The eq. wt. of H_2SO_4 = $\dfrac{\text{Mol. wt. of } H_2SO_4}{\text{No. of replaceable hydrogen atoms}}$

$= \dfrac{98}{2} = 49.$

STEP II

To prepare 1N H_2SO_4, we need 1×49 g

i.e., 49 g of H_2SO_4 for 1 litre of the solution.

To prepare 2N H_2SO_4, we need 2×49 g

i.e., 98 g of H_2SO_4 for 1 litre of the solution.

STEP III

Find the formula weight (or Molecular wt.) of H_2SO_4.

Formula (or Molecular) weight $= 2 + 32 + 64 = 98$.

STEP IV

To prepare 1F solution, we need 1×98 g of H_2SO_4

i.e., 98 g of H_2SO_4 for 1 litre of the solution.

To prepare 2F solution, we need 2×98 g of H_2SO_4

or 196 g of H_2SO_4 for 1 litre of the solution.

To prepare 1M solution, we need 1×98 g of H_2SO_4

or 98 g of H_2SO_4 for 1 litre.

To prepare $\dfrac{M}{2}$ solution we need $\dfrac{1}{2} \times 98$ g of H_2SO.

i.e., 49 g for 1 litre of the solution.

Example 3. *What is the normality of solution containing 5.3 g Na_2CO_3 in 200 ml ?*

STEP I

Find the strength per litre of Na_2CO_3 solution.

Now 200 ml solution of Na_2CO_3 contains $Na_2CO_3 = 5.3$ g

∴ 1000 ml solution of Na_2CO_3 contains Na_2CO_3

$$= \dfrac{5.3}{200} \times 1000 = 26.5 \text{ g}$$

STEP II

Find the gram-equivalent of Na_2CO_3.

Gram equivalent of any substance

$$= \dfrac{\text{Wt. in grams}}{\text{Eq. wt. of the substance}}$$

∴ Gram equivalent of Na_2CO_3

$$= \dfrac{26.5}{53.0} = \dfrac{1}{2} \text{ eqts.}$$

STEP III

Find the normality.

Normality $= \dfrac{\text{No. of gram-equivalent}}{\text{Volume in litres}}$

$$= \dfrac{\frac{1}{2}}{1} = \dfrac{1}{2}$$

Hence the normality of $Na_2CO_3 = \dfrac{N}{2}$.

Example 4. *Find the formality and molarity of $(NH_4)_2SO_4$ solution, 12 g of which have been dissolved in 250 ml solution.*

$$\text{Formality} = \frac{\text{No. of gram-formula weights}}{\text{Volume of the solution in litres}}$$

and $$\text{Molarity} = \frac{\text{No. of moles}}{\text{Volume of the solution in litres}}$$

STEP I

Find the number of gram formula weights and moles.

No. of gram-formula weights

$$= \frac{\text{Wt. of the substance in grams}}{\text{Formula weight}}$$

No. of gram-formula wt. of $(NH_4)_2SO_4 = \frac{12}{132}$

No. of moles of $(NH_4)_2SO_4 = \frac{12}{132}$

[here gram-formula wt. and molecular wt. of $(NH_4)_2SO_4$ are same.]

STEP II

Find the Formality or Molarity.

$$\text{Formality of } (NH_4)_2SO_4 = \frac{\frac{12}{132}}{\frac{250}{1000}} = \frac{12}{132} \times \frac{1000}{250}$$

$$= \frac{4}{11} \text{ F } \left(\text{or } \frac{4F}{11} \right)$$

Molarity of $(NH_4)_2SO_4$ is the same as the formality

i.e., $$\text{Molarity} = \frac{\frac{12}{132}}{\frac{250}{1000}} = \frac{12}{132} \times \frac{1000}{250} = \frac{4M}{11}$$

Example 5. *Calculate the molality of a solution which contains 10 g K_2CO_3 per 1000 g of water.*

Volumetric Analysis

STEP I

Find the moles of K_2CO_3.

$$\text{Moles of } K_2CO_3 = \frac{\text{Wt. of } K_2CO_3 \text{ in grams}}{\text{Molecular weight of } K_2CO_3}$$

$$= \frac{10}{138}$$

STEP II

Find the molality of K_2CO_3 solution.

$$\text{Molality} = \frac{\text{No. of moles}}{\text{Wt. of the solvent in kilograms}}$$

The weight of the solvent (water)

$$= 1000 \text{ grams}$$
$$= 1 \text{ kilogram}$$

$$\therefore \quad \text{molality} = \frac{\frac{10}{138}}{1} = \frac{10}{138} \text{ moles/kg.}$$

$$= 0.0724.$$

Example 6. *Find the normality, molarity and molality of a 10% solution of potassium hydroxide by weight. The resulting solution of potassium hydroxide has density 1.20 g/ml.*

10% solution of KOH by weight means 10 g of KOH are present in 100 g of the solution.

∴ wt. of water in 100 g of the solution

$$\therefore \quad = 100 - 10 = 90 \text{ g}$$

Now density of KOH solution is 1.20 g/ml which means that the weight of 1 ml solution of KOH is 1.20 g. This helps us in finding the volume of 100 g of KOH solution.

Now 1.20 g of KOH solution occupy volume = 1 ml

$$\therefore \quad 1 \text{ gm of KOH will occupy volume} = \frac{1}{1.20} \text{ ml.}$$

and 100 g KOH solution will occupy volume

$$= \frac{1}{1\cdot 20} \times 100$$
$$= 83\cdot 33 \text{ ml.}$$

STEP I

Find the normality.

$$\text{Normality} = \frac{\text{No. of g·equivalents}}{\text{Volume of solution in litres}}$$

$$= \frac{\dfrac{\text{Wt. in grams}}{\text{Eq. wt. of KOH}}}{\dfrac{\text{Vol. in ml.}}{1000}}$$

$$= \frac{\dfrac{10}{56}}{\dfrac{83\cdot 33}{1000}} = \frac{10}{56} \times \frac{1000}{83\cdot 33}$$

$$= 2\cdot 1$$

STEP II

Find the molarity.

$$\text{Molarity} = \frac{\text{No. of moles}}{\text{Volume of the solution in litres}}$$

$$= \frac{\dfrac{10}{56}}{\dfrac{83\cdot 33}{1000}} = 2\cdot 1$$

Since the eq. wt. and mol. wt. of KOH are same, the normality and molarity will also be same.

STEP III

Find the molality.

$$\text{Molality} = \frac{\text{No. of moles}}{\text{Wt. of the solution in kilograms}}$$

Volumetric Analysis

$$= \frac{\dfrac{\text{Wt. of KOH}}{\text{Mol. wt. of KOH}}}{\dfrac{\text{Wt. of water in grams}}{1000}}$$

$$= \frac{\dfrac{10}{56}}{\dfrac{90}{1000}} = \frac{10}{56} \times \frac{1000}{90}$$

$$= 1 \cdot 984 \text{ m (or } g \text{ moles/kilogram)}.$$

Example 7. *How many gram-equivalents of Ferrous sulphate are contained in (i) $0 \cdot 2$ N solution (ii) $3N$ solution (iii) $2 \cdot 5$ N solution ?*

Since 1N solution of any substance contains 1 gram-equivalent weight of it per litre.

∴ $0 \cdot 2$ N solution of $FeSO_4$ will contain $0 \cdot 2 \times 1$ g eq. $(0 \cdot 2 \times 152$ g) of $FeSO_4$ per litre.

Thus the number of gram equivalents of $FeSO_4$/litre in $0 \cdot 2$ N solute $= 0 \cdot 2$.

3 N solution of $FeSO_4$ will contain 3 gram-equivalents (or 3×152 g) of $FeSO_4$/litre.

$2 \cdot 5$ N solution of $FeSO_4$ will contain $2 \cdot 5$ g equivalents (or $2 \cdot 5 \times 152$ g) of $FeSO_4$/litre.

Example 8. *How many ml of $\dfrac{N}{10}$ solution can be prepared from $2 \cdot 86$ g $Na_2CO_3.10H_2O$?*

Eq. wt. of $Na_2CO_3.10H_2O = \dfrac{\text{Mol. wt.}}{2}$

$$= \frac{286}{2} = 143$$

Now 143 g $Na_2CO_3.10H_2O$ can give us solution

$$= 1000 \text{ ml 1N or } 10000 \text{ ml } \frac{N}{10}$$

∴ $2 \cdot 86$ g of $Na_2CO_3.10H_2O$ will give $= \dfrac{10000}{143} \times 2 \cdot 86$

$$= 200 \text{ ml}$$

NEUTRALIZATION

We know that 1N solution of any acid contains 1 gram-equivalent of the substance per litre. If we want to neutralize 1 litre of 1N acid, we shall need 1 litre of 1N solution of any base since 1 litre 1N solution of the base will also contain 1 g equivalent of the base dissolved in it. Thus, if we wish 1 N acid (1 litre) to be neutralized by 0.5 N solution of a base we will need 2 litres of this base. The comparison of the normalities of the acid and the base shows that the normality of the acid is double that of the base. Naturally to neutralize 1 litre 1N acid we require double the volume of the base *i.e.*, 2 litres. In other words we can say that the product of the normality of the acid and its volume is equal to the product of the normality of the base and its volume *i.e.*,

$$1 \times 1 = 0.5 \times 2$$

It has been found that just as in Bolye's Law

$$P_1 \times V_1 = P_2 \times V_2,$$

here in neutralization of an acid by a base also we have

Normality of the acid × Volume of the acid = Normlality of the base × Volume of the base

or $\qquad N_1 \times V_1 = N_2 \times V_2.$

This is called the *normality equation* in volumetric analysis.

Example 9. *What volume of 0.5N NaOH is required to neutralize 50 ml of 1.5N HCl ?*

Applying the normality equation and substituting the data, we have

$$\begin{array}{cc} acid & base \\ N_1 \times V_1 = & N_2 \times V_2 \\ 1.5 \times 50 = & 0.5 \times V_2 \end{array}$$

or $\qquad V_2 = \dfrac{1.5 \times 50}{0.5} = 150$ ml.

Example 10. *What is the normality of the acid solution 20 ml of which was neutralized by 25 ml of 0.5N NaOH solution ?*

Applying the normality equation and substituting the values we have

$$N_1 \times V_1 = N_2 \times V_2$$

or $\quad N_1 \times 20 = 0.5 \times 25$

or $\quad N_1 = \dfrac{0.5 \times 25}{20}$

$\quad\quad = 0.625$

Example 11. *Calculate the number of ml of $0.1\ N\ H_2SO_4$ solution necessary to react completely with a solution that contains $0.106\ g$ of pure Na_2CO_3.*

Step I

Find the gram equivalent of Na_2CO_3.

Gram-equivalents of Na_2CO_3 contained in 0.106 g.

$$= \dfrac{\text{wt. of } Na_2CO_3 \text{ in grams}}{\text{g eq. wt. of } Na_2CO_3}$$

$$= \dfrac{0.106}{53} = 0.002.$$

Step II

Find the ml of $0.1N\ H_2SO_4$ required.

According to the law of equivalents (substances react in the ratio of their equivalent weights) we can say that to neutralize 0.002 g eqts. of Na_2CO_3 we need the same amount *i.e.*, 0.002 g eqts. of the acid as well.

We know that normality $= \dfrac{\text{No. of gram equivalents}}{\text{Volume of the solution in litres}}$

Substituting the values of the given acid in this relation

$$0.1 = \dfrac{0.002}{V}$$

or $\quad V = \dfrac{0.002}{0.1} = 0.02$ litre $= 20$ ml

Thus, the volume of 0·1N acid required for the complete neutralization of 0·106 g of Na_2CO_3 is 20 ml.

Example 12. *It is found that 20·0 ml of H_2SO_4 solution reacted completely with 0·20 g of pure NaOH. What was the normality of the acid solution?*

STEP I

Find the gram equivalents of NaOH.

No. of gram-equivalents in a substance

$$= \frac{\text{wt. of the substance in g}}{\text{gram -eq. wt.}}$$

∴ No. of g equivalent in 0·20 g NaOH

$$= \frac{0\cdot 2}{40} = 0\cdot 05$$

STEP II

According to the law of equivalents, 0·05 g eqts. of NaOH must be neutralized with 0·05 g eqts. of the acid.

We know that normality $= \dfrac{\text{No. of g-equivalents}}{\text{Volume in litres}}$

∴ Normality of $H_2SO_4 = \dfrac{0\cdot 05}{\dfrac{20}{1000}} = 2\cdot 5$

DILUTION PROBLEMS

Example 13. *How many ml of 12N HCl solution is required for the preparation of 250 ml of $\dfrac{N}{10}$ HCl?*

Applying the normality equation

$$N_1 \times V_1 = N_2 \times V_2$$
$$N_1 = 12N$$
$$V_1 = ?$$
$$N_2 = \frac{N}{10}$$
$$V_2 = 250 \text{ ml.}$$

Substituting the values in the above equation we have

$$12 \times V_1 = \frac{1}{10} \times 250$$

or

$$V_1 = \frac{250}{10 \times 12} = 2 \cdot 083 \text{ ml.}$$

Example 14. *What shall be the normality of H_2SO_4 solution which is obtained by diluting 20 ml of $30N$ H_2SO_4 with 250 ml water?*

Now there are two acid solutions, one is $30N$ H_2SO_4 (Acid A) and the other is diluted with 250 ml water (Acid B).

Applying the normality equation,

$$\begin{array}{cc} Acid\ A & Acid\ B \\ N_1 \times V_1 & = N_2 \times V_2 \end{array}$$

$$N_1 = 30\ (N)$$
$$V_1 = 20\ \text{ml}$$
$$N_2 = ?$$
$$V_2 = 20\ \text{ml (acid of 30 N)} + 250\ \text{ml}$$
$$\text{(water)}$$
$$= 270$$

Substituting the values we have

$$30 \times 20 = N_2 \times 270$$

or

$$N_2 = \frac{30 \times 20}{270} = 2 \cdot 2.$$

Thus normality of the acid after adding 250 ml of water wil be $2 \cdot 2$ N.

Example 15. *What volume of $\frac{N}{3}$ HCl and $\frac{N}{12}$ HCl be mixed to get 1 litre of $\frac{N}{6}$ HCl?*

Let the volume of $\frac{N}{3}$ HCl requried $= x$ litres

∴ the volume of $\frac{N}{12}$ HCl required $= 1-x$ litres

We know that normality = $\dfrac{\text{g-equivalents}}{\text{vol. of the solution in litres}}$

$\therefore \quad \dfrac{1}{3} = \dfrac{\text{g-eqts.}}{x}$ in case of $\dfrac{N}{3}$ HCl

or g-equivalent of $\dfrac{N}{3}$ HCl $= \dfrac{1}{3} \times x = \dfrac{x}{3}$

Similarly g-equivalents of $\dfrac{N}{12}$ HCl $= \dfrac{1}{12} \times (1-x)$

and the g-equivalents of $\dfrac{N}{6}$ HCl $= 1 \times \dfrac{1}{6} = \dfrac{1}{6}$.

Thus $\dfrac{x}{3} + \dfrac{1}{12}(1-x) = \dfrac{1}{6}$

or $x = \tfrac{1}{2}$ litre

Thus volume of $\dfrac{N}{3}$ HCl $= 0.333$ litre

and volume of $\dfrac{N}{12}$ acid $= 1 - 0.333 = 0.667$ litre.

SPECIFIC GRAVITY OF SOLUTIONS

Example 16. *How many ml of 20% H_2SO_4 by weight (Specific gravity 1·14) is required for the neutratlization of 20 ml of 1·5N NaOH.*

STEP I

Find the gram equiavlent of NaOH present in 20·0 ml of 1·5N NaOH.

$$\text{Normality} = \dfrac{\text{gram-equivalent}}{\text{Volume of the solution in litres}}$$

$\therefore$ gram equivalents = Normality $\times$ Vol. in litres

$$= 1.5 \times \dfrac{20}{1000} = 0.03$$

STEP II

Find the g-eqts. of H_2SO_4.

To neutralize 0·03 gram equivalents of NaOH, we need 0·03 g. eqts. of H_2SO_4.

Volumetric Analysis

Now 1 g-equivalent of $H_2SO_4 = 49$ g.

∴ 0·03 g. eqts. of $H_2SO_4 = 0.03 \times 49 = 1.47$

Step III

Convert the weight of acid into 20% solution by weight.

20% H_2SO_4 means that in 100 g H_2SO_4 solution, 20 g of pure H_2SO_4 are present or we can say

that 20 g of pure H_2SO_4 can give us solution = 100 g

$$1 \quad ,, \quad ,, \quad ,, \quad ,, \quad ,, \quad = \frac{100}{20} \text{ g}$$

$$1.47 \text{ g} \quad ,, \quad ,, \quad ,, \quad ,, \quad ,, \quad = \frac{100}{20} \times 1.47$$

$$= 7.35 \text{ g}$$

Step IV

Convert the weight of H_2SO_4 acid solution into volume.

We are given that the specfic gravity of H_2SO_4 solution is 1·14.

This means that 1·14 g of H_2SO_4 solution occupy = 1 ml

$$1 \quad ,, \quad ,, \quad ,, \quad ,, \quad = \frac{1}{1.14} \text{ ml}$$

and $7.35 \quad ,, \quad ,, \quad ,, \quad ,, \quad = \frac{1}{1.14} \times 7.35$ ml

$$= 6.44 \text{ ml}.$$

Hence 6·44 ml of 20% H_2SO_4 solution are required to neutralize 20 ml of 1·5N NaOH.

DETERMINATION OF EQUIVALENT WEIGHTS, MOLECULAR WEIGHTS AND ATOMIC WEIGHTS

In all these problems, the relation

Strength = Normality × Equivalent weight

is made use of. The equivalent weight of the substance can be known if the strength per litre of its solution and its normality are known.

WHAT IS BASICITY ?

In case of acids we know that

$$\text{Eq. wt.} = \frac{\text{Molecular weight of the acid}}{\text{The no. of replaceable hydrogen atoms}}$$

The *number of replaceable hydrogen atoms present in an acid is called its basicity*. Naturally if out of three values, two are known, the third can be calculated. Thus we conclude that in case of acids.

$$\text{Eq. wt.} \times \text{Basicity} = \text{Molecular weight}$$

HCl is *mono-basic* because it has one replaceable hydrogen atom.

H_2SO_4 is *di-basic* because it has two replaceable hydrogen atoms.

WHAT IS ACIDITY ?

In case of bases, we know that

$$\text{Eq. wt.} = \frac{\text{Mol. wt. of the base}}{\text{No. of replaceable 'OH' groups}}$$

The no. of replaceable 'OH' groups in a base is called the acidity of the base. Thus NaOH, KOH are *mono-acid* bases since they contain only one 'OH' group in their molecules.

$Ca(OH)_2$, $Mg(OH)_2$ are di-acid bases since they have two 'OH' groups in their molecule.

Thus we find that in case of bases

$$\text{Mol wt.} = \text{Acidity} \times \text{Eq. wt.}$$

If any two values are known, the third can be found out.

The atomic weight of a metal can be determined by making use of the relation

$$\text{At. wt.} = \text{Valency} \times \text{Eq. wt.}$$

Example 17. *0·075 g of a mono basic acid required 10 ml of N/12 NaOH solution for complete neutralization. Calculate the molecular weight of the acid.*

STEP I

Find the no. of g-equivalents of NaOH.

Applying the relation, $\text{normality} = \dfrac{\text{No. of g-eqts.}}{\text{Volume in litres}}$

No. of g-eqts. = Normality × volume in litres

$$= \frac{1}{12} \times \frac{10}{1000} = \frac{1}{1200}$$

STEP II

To neutralize $\dfrac{1}{1200}$ g eqts. of NaOH, we need $\dfrac{1}{1200}$ g eqts. of the acid as well (on the principle of equivalency).

We know, that no. of g eqts. $= \dfrac{\text{wt. in g}}{\text{g. eq. wt.}}$

Let the eq. wt. of the acid $= E$

∴ No. of g eqts. of the acid in 0·075 g of it.

$$= \frac{0.075}{E}$$

∴ $\quad \dfrac{0.075}{E} = \dfrac{1}{1200}$

or $\quad\quad\quad\quad\quad E = 0.075 \times 1200 = 90.000$

Hence, the equivalent weight of the acid = 90.

Since $\quad\quad$ Mol. wt. = Basicity × Eq. wt.

Mol. wt. of the acid $= 1 \times 90 = 90$.

Example 18. *0·20 g of a di-acid base required 26 ml of N/10 HCl for complete neutralization. Calculate the molecular weight of the base.*

STEP I

Find the gram-equivalent of the acid.

We know that

$$\text{Normality} = \frac{\text{No. of g-equivalents}}{\text{volume of the solution in litres}}$$

or No. of g-equivalents = Normality × vol. in litres

$$= \frac{1}{10} \times \frac{25}{1000} = \frac{1}{400}$$

STEP II

To neutralize $\frac{1}{400}$ g. eqts. of the acid, we require $\frac{1}{400}$ g. eqts. of the base as well.

Let the Eq. wt. of the base $= E$

Since gram equivalent of a substance is

$$= \frac{\text{wt. of the substance in grams}}{\text{Eq. wt. of the substance}}$$

∴ g. eqts. of the base $= \frac{0\cdot 20}{E}$

Thus $\frac{0\cdot 20}{E} = \frac{1}{400}$

or $E = 0\cdot 20 \times 400 = 80\cdot 00$

∴ Molecular weight of the base $= 2 \times 80 = 160$.

Example 19. *5·3 g of a monovalent metal carbonate was dissolved per litre of the solution. 10 ml of this carbonate solution required 20 ml of N/20 HCl solution. Find the equivalent weight and atomic weight of the metal.*

STEP I

Find the normality of the metal carbonate solution.

Applying the normality equation :

$$N_1 \times V_1 = N_2 \times V_2$$

(*acid*)   (*metal carbonate*)

$$\frac{1}{20} \times 20 = N_2 \times 10$$

∴ $N_2 = \frac{1}{10}$,

Step II

Find the Eq. wt. of the metal carbonate.

We know that strength = Normality × Eq. wt.

∴ $5.3 = \dfrac{1}{10} \times$ Eq. wt.

∴ Eq. wt. of the metal carbonate = $5.3 \times 10 = 53$

Let this metal be M

Since the metal is monovalent, the formula of its carbonate is M_2CO_3

The equivalent weight of $M_2CO_3 = 53$

The equivalent weight of metal carbonate M_2CO_3 is the sum total of the equivalent weight of the metal and that of the carbonate (*i.e.*, eq. wt. of the metal + eq. wt. of CO_3^{-2}).

∴ Eq. wt. of the metal

= Eq. wt. of metal carbonate − Eq. wt. of CO_3^{-2}

Let the Eq. wt. of the metal = E.

and the Eq. wt. of $CO_3^{-2} = \dfrac{12 + 48}{2} = \dfrac{60}{2} = 30$

Thus, $E + 30 = 53$

∴ $E = 53 - 30 = 23.$

Thus, the Eq. wt. of the metal = **23**.

Step III

Find the atomic weight of the metal.

Since its valency = 1 (being monovalent)

and atomic weight = Eq. wt. × Valency

∴ At. wt. = $23 \times 1 = $ **23**.

Example 20. *20.0 ml of an acid containing 1.0 g dissolved in 500 ml solution required for a complete neutralization 19.15 ml of 0.122N NaOH. If the molecular weight of the acid be 74.6 find the basicity.*

Step I

Find the normality of the given acid.

Applying the normality equation,

$$N_1 \times V_1 = N_2 \times V_2$$
$$\text{(base)} \quad \text{(acid)}$$

$$0.112 \times 19.15 = N_2 \times 20$$

$$\therefore \quad N_2 = \frac{0.112 \times 19.15}{20} = 0.107$$

Step II

Find the strength per litre of the acid solution.

Since 2·0 g of acid are dissolved in 500 ml

∴ Strength per litre = 4·0 g.

Step III

Find the equivalent weight of the acid.

We know that Strength = N × Eq. wt.

or $\quad$ Eq. wt. $= \dfrac{\text{Strength}}{\text{Normality}} = \dfrac{4}{0.107} = 37.3$ g

Step IV

Find the basicity.

We know that Mol. wt. = Eq. wt. × basicity

$$\therefore \text{Basicity of the acid} = \frac{\text{Mol. wt.}}{\text{Eq. wt.}}$$

$$= \frac{74.6}{37.38} = 2.$$

COMPOSITION AND PERCENTAGE PROBLEMS

Example 21. *Exactly 20 ml of 0·8N acid was required to neutralize completely 1·12 g of an impure sample of calcium oxide. Calculate the percentage purity of the calcium oxide.*

Step I

Find the gram-equivalents of the acid.

We know that $\text{normality} = \dfrac{\text{g-equivalents of acid}}{\text{volume of acid in litres}}$

or   gram equivalents of HCl = Normality × Vol. in litre

$$= 0.8 \times \frac{20}{1000} = 0.016$$

STEP II

Since substances react in the ratio of their equivalent weights, 0.016 g eqts. of the acid will neutralize 0.016 g eqts. of CaO.

Now 1 g equivalent of CaO = $\dfrac{\text{Mol. wt. of CaO}}{2}$

$$= \frac{40+16}{2} = 28$$

∴   0.016 g eqts. of CaO = 28 × 0.016 = 0.448 g.

STEP III

Thus 1.12 g of impure CaO contains pure CaO
$$= 0.448 \text{ g}$$

or 1 g of impure CaO will contain = $\dfrac{0.448}{1.12}$ g of pure CaO

∴   % of CaO   = $\dfrac{0.448}{1.12} \times 100 = 40$

Thus percentage purity of CaO = 40.

Example 22. *What is the purity of concentrated sulphuric acid (Specific gravity = 1.8) if 5 ml is neutralised by 84.0 ml of 2N NaOH ?*

STEP I

Find the gram-equivalents of NaOH

Applying the relation

$$\text{normality} = \frac{\text{gram-equivalents}}{\text{volume of solution in litres}}$$

or   No. of g eqts. of NaOH = Normality × Vol. in litres

$$= 2 \times \frac{84}{1000} = 0.168.$$

Step II

Find the weight of 5 ml of H_2SO_4.

The specific gravity of H_2SO_4 = 1·8

It means that 1 ml of H_2SO_4 solution weights = 1·8 g

∴ 5 ml of H_2SO_4 solution weighs = $5 \times 1\cdot8$ = 9·00 g

Step III

Find the weight of pure H_2SO_4 in 9·00 g of impure specimen.

Now 0·168 g eqt. of NaOH must be neutralised by 0·168 g eqts. of H_2SO_4 according to the law of equivalents.

Now 1 g. eqt. of H_2SO_4 = 49 g $\left(\dfrac{\text{Mol. wt.}}{2}\right)$

∴ 0·168 g eqt. of H_2SO_4 = $0\cdot168 \times 49$ = 8·232 g.

Step IV

Find the percentage purity of H_2SO_4.

9·00 g of impure H_2SO_4 contains pure H_2SO_4 = 8·232 g

1 g of impure H_2SO_4 contains pure H_2SO_4 = $\dfrac{8\cdot232}{9\cdot00}$

100 g impure H_2SO_4 contains pure H_2SO_4

$$= \dfrac{8\cdot232}{9\cdot0} \times 100$$

$$= \dfrac{823\cdot2}{9} = 91\cdot46 \text{ g}.$$

Thus percentage purity of H_2SO_4 = 91·46.

Example 23. *Exactly 50 ml sample of sodium carbonate was titrated with 60·0 ml of 3N HCl. If the specific gravity of the sodium carbonate solution is 1·25, what percentage by weight of Na_2CO_3 does it contain?*

Step I

Find the no. of gram-equivalents of 3N HCl.

We know that

$$\text{Normality} = \dfrac{\text{No. of g-equivalents}}{\text{Vol. of the solution in litres}}$$

Volumetric Analysis

or no. of g equivalents = Normality × Vol. in litres

$$= 3 \times \frac{60}{1000} = 0.18 \text{ g eqt. of HCl.}$$

STEP II

Find the weight of 50 ml of Na_2CO_3 solution.

We know that the specific gravity of sodium carbonate solution is 1·25. This means that 1 ml of this solution weighs 1·25 g.

Thus 1 ml of Na_2CO_3 solution weighs = 1·25 g

50 ml of Na_2CO_3 solution weighs = 1·25 × 50 = 62·50 g

STEP III

Now 0·18 g eqt. of HCl will be neutralized by 0·18 g eqt. of Na_2CO_3 according to the law of equivalents.

Now 1 g equivalent of $Na_2CO_3 = \dfrac{\text{Mol. wt. of } Na_2CO_3}{2}$

$$= \frac{106}{2}$$

$$= 53$$

∴ 0·16 g eqt. of Na_2CO_3

$$= 53 \times 0.18$$

$$= 9.54 \text{ g}$$

STEP IV

Find the percentage of Na_2CO_3.

The total weight of Na_2CO_3 solution

$$= 62.5 \text{ g}$$

Now 62·5 g of the solution contains pure Na_2CO_3

$$= 9.54 \text{ g}$$

1 g of the solution contains pure $Na_2CO_3 = \dfrac{9.54}{62.5}$

100 g of the solution contains pure $Na_2CO_3 = \dfrac{9.54}{62.5} \times 100$

$$= 15.26$$

Hence the given Na_2CO_3 solution is 15·26% by weight.

Example 24. 3·0 g of a mixture of NaCl and NaOH were dissolved in distilled water to make 250 cc of the solution. 20 cc of this solution required 24 cc HCl for neutralization. 18 cc of this HCl required 20 cc of N/10 KOH solution for complete neutralization. Find the percentage composition of the mixture.

STEP I

Find the Normality of the HCl solution.

$$N_1 \times V_1 = N_2 \times V_2$$
$$(acid) \quad (base)$$

Substituting the values, we have

$$N_1 \times 18 = \frac{1}{10} \times 20$$

∴ $$N_1 = \frac{20}{180} = \frac{1}{9}$$

STEP II

Find the normality of the mixture with respect to NaOH (since out of NaCl and NaOH only NaOH reacts with HCl),

$$N_1 \times V_1 = N_2 \times V_2$$
$$(acid) \quad (mixture)$$

$$\frac{1}{9} \times 24 = N_2 \times 20$$

$$N_2 = \frac{24}{9 \times 20} = \frac{2}{15}$$

STEP III

Find the strength per litre of NaOH,

Applying the relation

$$\text{Strength} = \text{Normality} \times \text{eq. wt.}$$

and substituting the values (for NaOH) we have

$$\text{Strength} = \frac{2}{15} \times 40 \text{ (eq. wt. of NaOH)}$$

$$= 5·33 \text{ g/litre}$$

STEP IV

Find the strength of the given mixture solution/litre.

We are given that 250 cc of mixture solution contains mixture = 3·0 g

∴ 1000 cc (one litre) will contain = 12 g

STEP V

Find the per cent of each component in the mixture.

Now 12 g of the total mixture contains NaOH = 5·33 g

∴ 1 g of the total mixture contains NaOH = $\frac{5·33}{12}$ g and 100 g of the total mixture contains NaOH

$$= \frac{5·33}{12} \times 100 = 44·41 \text{ g}$$

Thus, the percentage of NaOH = 44·41

and the percentage of NaCl = 100 − 44·41 = 55·59

Example 25. *6·50 g of a mixture of carbonates of sodium and potassium are dissolved per litre of the solution. 20 ml of the solution required 22·0 ml of 0·101N H_2SO_4 for complete neutralization. Calculate the percentage composition of the mixture.*

Here both the components of the mixture react with the acid.

STEP I

Find the normality of the mixture solution.

$$N_1 \times V_1 = N_2 \times V_2$$
$$(acid) \quad (mixture)$$

$$0·101 \times 22 = N_2 \times 20$$

∴ $N_2 = \dfrac{0·101 \times 22}{20} = 0·111$ (total normality of both the components)

STEP II

Find the theoretical normalities of each of the two components

Let the amount of Na_2CO_3 in the mixture $= x$ g

∴ the amount of K_2CO_3 in the mixture $= 6·5 - x$ g

We know that

$$\text{Strength} = \text{Normality} \times \text{Eq. wt.}$$

or $\quad \text{Normality} = \dfrac{\text{Strength}}{\text{Eq. wt.}}$

∴ The normality of $Na_2CO_3 = \dfrac{x}{53 \text{(Eq. wt. of } Na_2CO_3)}$

and the normality of $K_2CO_3 = \dfrac{6·5 - x}{69 \text{ (Eq. wt. of } K_2CO_3)}$

Total theoretical normality of both $= \dfrac{x}{53} + \dfrac{6·5 - x}{69}$

STEP III

Find the value of x.

Now total (observed) normality $= 0·111$ N

and total theoretical normality $= \dfrac{x}{53} + \dfrac{6·5 - x}{69}$

∴ $\quad \dfrac{x}{53} + \dfrac{6·5 - x}{69} = 0·111$

Solving the above equation we have

$$x = 3·839$$

i. e., the amount of $Na_2CO_3 = 3·839$ g

and that of $K_3CO_3 = 6·5 - 3·839 = 2·661$ g

STEP IV

Find the per cent of each.

Now 6·5 g of the mixture contains $Na_2CO_3 = 3·839$ g

1 g of the mixture contains $Na_2CO_3 = \dfrac{3·839}{6·5}$ g

Volumetric Analysis

100 g of the mixture contains $Na_2CO_3 = \dfrac{3.839}{6.5} \times 100$

$ = 59.06$ g

Thus the percentage of $Na_2CO_3 = 59.06$
and that of $K_2CO_3 = 100 - 59.06 = 40.94$.

WATER OF CRYSTALLISATION

Example 26. *12.6 g of hydrated oxalic acid,*

$$\left\{ \begin{array}{l} COOH \\ | \\ COOH \end{array} . xH_2O \right\}$$

was dissolved per litre of the solution. 10 ml of the solution required 20 ml of N/10 NaOH for complete neutralization. Find the value of x (no. of water molecules).

Step I

Find the normality of the acid solution.

Applying the equation

$$\begin{array}{cc} N_1 \times V_1 = N_2 \times V_2 \\ (acid) \quad (base) \end{array}$$

$$N_1 \times 10 = \dfrac{1}{10} \times 20$$

$$N_1 = \dfrac{20}{1000} = \dfrac{1}{5}$$

Step II

Find the strength of anhydrous oxalic acid.

The eq. wt. of anhydrous oxalic acid $= \dfrac{\text{Mol. wt.}}{2}$

$$= \dfrac{90}{2} (C_2H_2O_4) = 45$$

Applying the relation

Strength = Normality × Eq. wt.

and substituting the values, we have

$$\text{Strength} = \frac{1}{5} \times 45 = 9.0 \text{ g.}$$

STEP III

Find the value of x.

Applying the relation

$$\frac{\text{Mol. wt. of hydrated compound}}{\text{Mol. wt. of anhydrous compound}} = \frac{\text{wt. of hydrated compound}}{\text{wt. of anhydrous compound}}$$

and substituting the values, we have

Mol. wt. of hydrated oxalic acid $\left\{ \begin{array}{c} \text{COOH} \\ | \\ \text{COOH} \end{array} . x\text{H}_2\text{O} \right\}$

$$= 90 + 18x$$

Mol. wt. of anhydrous oxalic acid $\left\{ \begin{array}{c} \text{COOH} \\ | \\ \text{COOH} \end{array} \right\}$ 90

Thus

$$\frac{90 + 18x}{90} = \frac{12 \cdot 6}{9 \cdot 00}$$

Solving the above equation

$$9 (90 + 18x) = 90 \times 12 \cdot 6$$

or
$$x = 2$$

Thus, the no. of water molecules = 2

and the formula of hydrated oxalic acid is $\begin{array}{c} \text{COOH} \\ | \\ \text{COOH} \end{array}$ .2H$_2$O

End-of-chapter problems

Concentration

1. Find the number of gram equivalents or molecules of each substance in

(a) 500 ml of 0·5N NaOH

(b) 250 ml of 0·25N H$_2$SO$_4$

(c) 250 ml of 0·5M MgSO$_4$

(d) 1 litre of $\frac{M}{10}$ Na$_2$CO$_3$

(e) 250 ml of $\frac{N}{3}$ HCl

[**Ans.** (a) = 0.25 g.eqt.; (b) = 0.0625 g.eqt.; (c) = 0.125 g.moles;
(d) = 0.1 mole; (e) = 0.083 g.eqt.]

2. Calculate the normalities of the following :
(a) 0.4 g of NaOH dissolved in 250 ml solution
(b) 5.3 g of Na_2CO_3 dissolved in 250 ml solution
(c) 3.16 g of $KMnO_4$ dissolved in 100 ml solution
(d) 1.52 g of $FeSO_4$ dissolved in 250 ml solution
(e) 3.65 g of HCl dissolved in 500 ml solution
(f) 4.9 g of H_2SO_4 dissolved in 500 ml solution.

$$\left[\textbf{Ans.} \ (a) = \frac{N}{25}, \ (b) = \frac{2N}{5}, \ (c) = 1N, \ (d) = \frac{N}{25}, \ (e) = \frac{N}{5}, \ (f) = \frac{N}{5} \right]$$

3. Determine the molarity of the following solutions :
(a) 4.9 g of H_2SO_4 dissolved in 250 ml solution
(b) 8.55 g of $Al_2(SO_4)_3$ dissolved in 250 ml solution
(c) 104 g of $BaCl_2$ in 1 litre solution
(d) 73 g of HCl in 2 litre solution
(e) 286 g of $Na_2CO_3 \cdot 10H_2O$ in 2 litre solution.

$$\left[\textbf{Ans.} \ (a) = \frac{M}{5}, \ (b) = 0.05M, \ (c) = 0.5M, \ (d) = 1M, \ (e) = 0.5M \right]$$

4. What is the molality of the following solutions :
(a) 10 g of $C_{12}H_{22}O_{11}$ dissolved in 200 g of water
(b) 20 g of NaOH dissolved in 200 g of water
(c) 15.2 g. of $FeSO_4$ dissolved in 250 g of water
(d) 20.8 of $BaCl_2$ dissolved in 500 g of water.

$$\left[\textbf{Ans.} \ (a) = \frac{10}{81} m, \ (b) = \frac{1}{4} m, \ (c) = \frac{2}{5} m, \ (d) = \frac{1}{5} m. \right]$$

5. Determine the number of grams of the substance required to prepare :

(i) $\frac{N}{10}$ $KMnO_4$ (ii) $\frac{N}{2}$ NaOH, (iii) 0.1 $FeSO_4$,

(iv) $0.2N\ Na_2CO_3$, (v) $0.5N\ HS_2O_4$, (vi) $1N\ HCl$.

[**Ans.** (i) 3·16 g/litre, (ii) 20·0 g/litre,
(iii) 15·2 g anhyd. $FeSO_4$/litre,
(iv) 10·6 g/litre,
(v) 24·5 g litre, (vi) 36·5 g litre]

6. What is the weight of potassium permanganate present in 200 ml of $0.1N\ KMnO_4$?

[**Ans.** 0·632 g]

7. (i) Calculate the normality of a solution containing 0·53 gm of sodium carbonate per litre.

(ii) Calculate the mole fraction of sodium hydroxide in a solution containing 20 gm of sodium hydroxide in 81 gm of water.

(*Marathwada P. U. C., 1973*)

[**Ans.** (i) 0·1 (ii) 0·1]

8. (a) Calculate the amount in grams of a solute in the following solutions :

(i) 3·5 litres of 0·1M solution of KOH.

(ii) 750 ml of 0·5 molar solution of NaCl.

(b) Calculate the molarity of the following solutions :

(i) solution containing 36·5 gm of HCl in 125 ml of solution.

(ii) solution containing 8 grams of $AgNO_3$ per litre,

(*Punjab Pre-Univ., 1973*)

[**Ans.** (a) (i) 19·6 gm. (ii) 21·94 gm.
(b) (i) 8 (ii) 0·047]

9. Write directions to prepare the following solutions :

(i) 250 ml of $0.1N$-Na_2CO_3 solution.

(ii) 200 ml of ½ Molar KOH solution. (*Punjab H.S. 1975*)

[**Ans.** (i) Dissolve 1·325 gm· of solid Na_2CO_3 in water and make the vol. 250 ml.

(ii) Dissolve 5·6 gm. of solid KOH in water and make the vol. 200 ml.]

10. How will you prepare 1 m (molal) solution of sodium hydroxide ? How many hydroxyl ions will it contain.

(*Guru Nanak Dev Pre-Univ., 1978*)

[**Ans.** By dissolving 40 g of NaOH in 1000 g of H_2O ;
No. of Hydroxyl ions $= 6.023 \times 10^{23}$]

Dilution

11. Calculate the ml of water that must be added to 250 ml of 1·25 N HCl to make it 0·5N HCl. [**Ans.** 375 ml]

12. What volume of water is needed to prepare 9N H_2SO_4 from 9 ml of 36 N H_2SO_4 ? [**Ans.** 27 ml]

13. 50 ml of water is added to 10 ml of 6N HNO_3 solution. that is the normality of the resulting solution ? [**Ans.** 1]

14. How much volume of N H_2SO_4 is needed to convert it to 32·8 ml of 0·145N H_2SO_4 ?

15. A solution of Na_2CO_3 is prepared by dissolving exactly 6 gm of it in a litre of water. 25 ml of this solution require 26 ml of dilute H_2SO_4 solution for neutralisation. Calculate strength of the acid and normality. What volume of water is to be added to one litre of acid to make it exactly decinormal ? *(Calcutta Pre-Univ., 1973)*

[**Ans.** 5·33 gm/litre ; 0·1089 ; 81·8 ml]

Neutralization

16. What volume of 0·1 N acid is required to neutralize 500 ml of 0·1N solution of a base. [**Ans.** 500 ml]

17. What volume of 0·1N base is required to neutralize (a) 1 litre of 0·01 N acid, (b) 0·05N acid 1 litre).

[**Ans.** (a) = 100 ml (b) = 500 ml]

18. What is the normality of an acid solution, 11·25 ml of which was neutralized by 13·7 ml of 0·54N NaOH solution ? [**Ans.** 0·66N]

19. 25 ml of HCl solution containing 7·3 g of the acid per litre neutralized 30 ml of NaOH solution. 20 ml of this alkali solution was neutralized with 24 ml of H_2SO_4 solution. Calculate the normality and strength of sulphuric acid. [**Ans.** 0·1388N H_2SO_4 ; 6·8 gm litre]

20. What is the normality of a KOH solution 30 ml of which neutralized 31·8 ml of 0·1198N H_2SO_4 ? [**Ans.** 0·127N]

21. 10·0 ml of H_2SO_4 solution reacts with 0·265 g of pure Na_2CO_3. Calculate the normality of the acid. [**Ans.** 0·5N]

22. 20 ml of a solution containing 6·5 g of a dibasic acid per litre neutralized 21·5 ml of a solution of sodium hydroxide, 20 ml of N/20 HCl neutralized 22·18 ml of the same NaOH solution. Calculate the molecular weight of the acid.

(Na=23, H=1, C=12, O=16)

[**Ans.** 251·9]

23. 10 cc of sulphuric acid (sq. gravity 1·85) were diluted with distilled water to 1 litre 10 cc of this solution required 34·65 cc decinormal NaOH solution for the complete neutralization. Calculate the percentage by weight of H_2SO_4 in the original acid.

[**Ans.** 91·78]

24. How much volume of $\frac{M}{10}$ hydrochloric acid is required for complete reaction with 20 ml of $\frac{M}{20}$ magnesium hydroxide ? Write an equation for this reaction. *(Punjab Pre-Univ, 1973)*

[**Ans.** 20 ml; $Mg(OH)_2 + 2HCl \longrightarrow MgCl_2 + 2H_2O$]

Equivalent Weight, Molecular Weght

25. 0·230 g of a monobasic acid required 50 ml of N/10 NaOH solution for complete neutralization. Calculate equivalent weight and molecular weight of the acid. [Ans. eq. wt. and mol. wt.=46]

26. 0·225 g of a dibasic acid required 10·84 ml of 0·25N NaOH for complete neutralization. What is the molecular weight of the acid ? [Ans. 45]

27. 1·5 g of an organic acid was dissolved in water and volume made up to 300 ml. 10 ml of this acid required 12·3 ml of N/10 NaOH for complete neutralization. If the molecular weight of the acid be 121·9, find the basicity of this acid. [Ans. 3]

28. 0·10 g of a base required 25 ml of N 16 HCl for complete neutralization. If the molecular weight of the base is 128, find the acidity of the base. [Ans. 2]

29. 0·075 g of a monobasic acid required 10 ml of N/12 NaOH solution for complete neutralization. Calculate the molecular weight of the acid. [Ans. 90]

30. 0·5 g zinc was dissolved in 12·5 cc of 2·432N H_2SO_4. The excess acid required 21·68 cc of 0·0974N NaOH for the complete neutralization. Find the eq. wt. of zinc. [Ans. 32·5]

31. 25·0 ml of a solution of sodium carbonate requiqed 17·5 of 0·1N hydrochloric acid for exact neutralization. Calculate the normality of sodium carbonate solution. *(Bangalore XI Standard, 1969)* [Ans. 0·06]

32. 0·40 of a substance is dissolved in water and the volume is made up to 10 ccs the resulting solution was found to be 0·1N. What is the equivalent weight of the substance ?

(Behrampur Pre-Univ, 1969) [Ans. 40]

33. 125 cc of seminormal hydrochloric acid exactly neutralized 120·0 cc of a solution of a base containing 4·50 g per litre. Find the equivalent weight of the base.

(Punjabi Pre-Univ, 1980) [Ans. 8·64]

34. 0·075 g of an acid required 10 ml of N/12 NaOH solution for complete neuralization. Calculate the equivalent weight of the acid.

(Panjab Pre/Univ. 1980) [Ans. 90]

35. 1·20 g of NaCl was disseoved in H_2O and the solution of silver nitrate treated with excess of $AgNO_3$. The weight of the precipitated silver chloride after washing and drying was 2·94 g. Calculate the equivalent weight of sodium. (At. Wts. Ag=108, Cl=35·5)

(Himachal Pre-Univ, 1980) [Ans. 23·07]

36. Molecular weight of an acid is 98 and its basicity is 3. Calculate its equivalent weight

(*Punjab School Education Board XI, 1981*)

[Ans. approximately 33]

Percentage Purity

37. What is the purity of concentrated H_2SO_4 (specific gravity 1·8) if 15 ml is neutralized by 43·6 ml 12N·NaOH ?   [Ans. 92·2%]

38. The density of a commercial sample of H_2SO_4 is 1·8 ml. 10 ml of this acid was made up to one litre with water. 20 ml of this dilute acid required 60 ml of N/10 NaOH for neutralization. Calculate the percentage purity (by weight) of the commercial sample.

[Ans. 81.67%]

39. The density of a commercial sample of sulphuric acid was 1·8 g/cc. 10 cc of this acid was dissolved in water and the solution was made up to one litre. 20 cc of this solution required 60 cc of N/10 caustic soda solution for complete neutralization. Calculate the percentage purity of the given sample of sulphuric acid

[Ans. 81.66]

40. 1·3 gm of and impure sample of ammonium chloride was heated with an excess of caustic soda and the ammonia evolved was absorbed in 60 ml of $N/2H_2SO_4$ solution, the excess of acid required 1·4 ml of N NaOH for neutralization. Calculate the percentage purity of original sample.   *U. P. Board Inter,1971*)

[Ans. 65·93%]

41. 1·0 gm of a sample of sodium carbonate was dissolved in 150 ml of water. 24·15 ml of N HCl solution was added to it. Finally the mixture required 60 cc of ·11N sodium hydroxide for complete neutralization. Calculate the percentage of sodium carbonate in the given sample.   (*Ranchi Pre-Univ , 1971*)

[Ans. 93·01%]

42. A sample of Washing Soda contains $Na_2CO_3=60\%$; $NaHCO_3=20\%$; $NaCl=20\%$. What volume of carbon dioxide gas measured at 27 C and 700 mm pressure would be produced by treating 10 g of the sample with dilute HCl ?

[$Na=23, C=12, O=16, H=1$]

(*Haryana Board Hr. Sec, 1979*)

[Ans. 2·147 litres]

Water of Crystallisation

43. 0·787 g of hydrated oxalic acid $(COOH)_2 \cdot x\, H_2O$ were dissolved in complete and volume made up to 250 ml of this solution required for complete neutralization 18 ml of N/12 KOH. Calculate the value of x.   [Ans. 2]

44. 2·85 g of hydrated sodium carbonate Na_2CO_3x/H_2O was dissolved in 100 ml solution. 10 ml of this solution required 20 ml of N/20 NaOH for complete neutralization. Find the number of water molecules in Na_2CO_3.   [Ans. 10]

45. 0·100 g of anhydrous organic acid require 19·84 ml of 0·112N NaOH for neutralization 0·250 g of the hydrated acid required 35·4 ml of the same alkal. Calculate the number of molecules of water of crystallisation per equivalent of the anhydrous acid [Ans 1]

46. 4·9 g of Mohr's salt are dissolved in water and the volume made up to 250 ml. 10 ml of this solution required 7·5 ml of N15 $KMnO_4$ for complete neutralization. Calulate the number of molecules, of water of crystallization in Mohr's salt. $[FeSO_4 \cdot (NO_4)_2SO_4 \cdot xH_2O]$.
[Ans. 6]

47. 1·245 gm of a sample of $CuSO_4 \cdot xH_2O$ was dissolved in water and H_2S passed till CuS was completely precipitated. The filtrate contained librated H_2SO_4 which required 20 ml of N/2 NaOH for neutralization. Calculate x, the number of molecules of H_2O associated with $CuSO_4$ (*U. P. Board Inter, 1971*)
[Ans. 5]

Percentage Composition

48. 1·5 g of a mixture of NaCl and NaOH were dissolved in distilled water to make 250 ml of the solution. 20 ml of this solution required 12 ml of HCl for neutralization and 9 ml of this acid required for complete neutralization 10 ml of N/10 KOH solution. Find the percentage of the mixture. [Ans. NaOH=44·44%]
[NaCl=55·56%]

49. 1·325 g of a mixture of Na_2CO_3 and NaCl was dissolved in water and volume raised to 250 ml of this solution was completely neutralized by 15 ml of N/15 KOH. Find the percentage composition of Na_2CO_3 and NaCl in the given mixture. [Ans Na_2CO_3=66·25%
NaCl=33·75%]

50. 9·6 g of a mixture of NaOH and KOH was dissolved in water and the volume made up to 1 litre. 5 ml of this mixture solution required for complete neutralization 15 ml of N 15 HCl. Calcu'ate the percentage of each component in the mixture. [Ans. NaOH=62·5%
[KOH=38·5%]

Miscellaneous

51. If ·10 ml of 0·45N NaOH solution is added to 15 ml of 0·32N HCl, is the resulting solution acidic or basic ? [Ans. Acidic]

52. 25 cc of dilute HCl liberate 10 cc of CO_2 at N. T. P. when treated with exeess of pure $CaCO_3$ calculate the normality of the acid
[Ans. 45N]

53. 1·06 g of anhydrous Na_2CG_3 is completely neutralized by 200 ml of H_2SO_4 solution. Calculate the normality and grams per litre of H_2SO_4. [Ans. 0·1, 4·9g l]

54. What volume of water should be added to 10 ml of a 2N solution of sulphiric acid to make it 0 4N ? [Ans· 40 ml]

55· Define molarity. Calculate molarity of hydrogen chloride in a solution when 0·365 g of it have been dissolved in 100 ml of the solution. (*Kurukshetra Pre-Univ., 1975*)
[Ans. 0·1M]

56. A sulphuric acid solution contains 53·0% sulphuric acid by weight and has a density of 1·835 g/ml. Calculate its molality and molarity.
(Gujarat Pre-Univ 1975)
[Ans. 18·98 m ; 34·83M]

57. How many grams of pure sulphuric acid will be required to prepare 100 ml of decinormal sulphuric acid solution ?
(At. wt. of sulphur 32)
(Gauhati Pre-Univ., 1976)
[Ans. 0·49 g]

58. How many molecules of sulphuric acid will be required to prepare 300 ml of 0·1M H_2SO_4 and 50 ml of 0·1N H_2SO_4 ?
(Maharaja Sayaji Rao Pre-Univ., 1976)
Ans. $6·023 \times 10^{22}$
$1·506 \times 10^{22}$]

59. You are supplied with 10 ml of N/2 solution of sodium hydroxide. How much water will you add to it to obtain N/10 sodium hydroxide solution ?
(All India Hr. Sec. 1977)
[Ans. 40 m]

60. Calculate the normality of a solution of NaOH when 0·4 g of it is dissolved in 100 ml of the solution.
(Himachal Pre-Univ., 1978)

$$\left[\text{Ans } \frac{3}{10} \right]$$

61. 0·637 gm. of acid required 21·6 ml. of seminormal NaOH for complete neutralisation. Calculate the equivalent weight of acid.
(Punjab Pre-Univ., 1982)
[Ans. 58·98]

62. 25 ml of a decinormal solution of an acid neutralises exactly 20·00 ml of the solution of a base, containing 2·40 gm of the base dissolved per 500 ml of the solution. Calculate the equivalent weight of the base.
(Guru Nanak Dev Pre-Univ., 1982)
[Ans. 38·4]

13

Problems Based on Equations

TYPE I—(Weight-Weight Relationship)

Much of the descriptive and quantitative knowledge about general Chemistry is condensed in the chemical equations. The formulae of the reacting substances represent the number of molecules which enter into combination with each other and the proportion by weight in which they react. Similarly, the formulae of the products formed in a reaction represents the number of molecules of each product formed and the ratio by weight in which these are formed. According to the **Law of Conservation of Mass**, *the total mass of the reactant must be equal to the total mass of the products formed. In fact, a* **balanced chemical equation** *is an algebraic equation with reactants on the left hand side (L.H.S) and the products on the right hand side (R.H.S.).* The coefficient of the formulae represents the number of molecules of various reactants and products in the chemical equation. For example in reaction

$$N_2 + 3H_2 \longrightarrow 2NH_3$$

There is no coefficient with N_2 which means that 1 molecule of nitrogen reacts with 3 molecules of hydrogen to give 2 molecules of ammonia. This proportion is true in case of weights as well as volumes of the reactants and the products, since all atoms and molecules have definite weights and volumes. It should be noted that the weight proportions may be expressed in any unit *e.g.*, grams, kilograms, pounds etc.

We shall now take up problems which are based on the weight relationship of the reactants and the products. It

Problems Based on Equations

should be noted that in these problems it is not necessary to know the conditions of temperature, pressure or concentration of the reactants or the products.

Points to be noted before solving these problems are :

(i) *Read the problem carefully.*

(ii) *Frame the balanced chemical equation involved in the problem if not given.*

(iii) *Mark the compounds actually involved in the problem.*

(iv) *Arrange the matter or the data in accordance with the suitability of the problem.*

Example 1. *Calculate the weight of mercuric oxide which is required to furnish 10.0 g of oxygen.*

Step I

Frame the chemical equation.

$$2HgO \longrightarrow 2Hg + O_2$$

Step II

Mark the compounds actually involved in the problem.

The compounds involved in the problem are mercuric oxide and oxygen. We have nothing to do with mercury even though in occurs in the reaction.

Step III

Find the molecular weights of the compounds actually taking part in the reaction.

The weight of HgO = atomic wt. of Hg + at. wt. of oxygen

$$= 200 \cdot 60 + 16 \cdot 00$$
$$= 216 \cdot 60$$

The weight of O_2 = 2 × at. wt. of oxygen

$$= 2 \times 16 = 32 \cdot 0$$

Step IV

Find the weight of mercuric oxide.

We see from the equation that

32 g of oxygen is obtained from HgO

$$= 433 \cdot 2$$

1 g of oxygen is obtained from HgO

$$= \frac{432 \cdot 2}{32}$$

∴ 10 g of oxygen will be obtained from HgO

$$= \frac{433 \cdot 2}{32} \times 10$$

$$= 135 \cdot 3 \text{ g}.$$

Example 2. *Calculate the weight of magnesium oxide (MgO) that can be obtained from 200 g of magnesium carbonate ($MgCO_3$) which is 95% pure.*

STEP I

Write the balanced chemical equation involved.

$$MgCO_3 \longrightarrow MgO + CO_2$$

STEP II

Mark the chemical compounds in the reaction and find their molecular weights.

The compounds involved in the problem are $MgCO_3$ and MgO.

Molecular weight of $MgCO_3 = 24 + 12 + 48$
$$= 84$$

Molecular weight of $MgO = 24 + 16 = 40$.

STEP III

Find the weight of pure $MgCO_3$.

It is given that $MgCO_3$ is 95% pure.

This means that in 100 g of impure sample of $MgCO_3$, pure $MgCO_3 = 95$ g.

So that the amount of pure $MgCO_3$ in 200 g of impure sample

$$= \frac{95}{100} \times 200$$

$$= 190 \text{ g}.$$

STEP IV

Find the amount of MgO obtained from $MgCO_3$

Problems Based on Equations

Now from the equation we see that

84 g of $MgCO_3$ give $MgO = 40$ g

1 g of $MgCO_3$ give $MgO = \dfrac{40}{84}$ g

190 g of $MgCO_3$ give $MgO = \dfrac{40}{84} \times 190$ g

$= 90.47$ g.

Example 3. *How many grams of 83.4% pure sodium sulphate can be produced from 250 g of 95% pure NaCl?*

Step I

Write the chemical equation involved:

$$2NaCl + H_2SO_4 \longrightarrow Na_2SO_4 + 2HCl$$

Step II

Find the molecular weight of the chemical compounds involved.

The chemical compounds involved in the problem are $NaCl$ and Na_2SO_4

Molecular weight of $NaCl = 23 + 35.5$
$= 58.5$

Molecular weight of $Na_2SO_4 = 46 + 32 + 64$
$= 142.$

Step III

Find the amount of pure NaCl in 250 g of impure sample.

Since NaCl is 95% pure, it means that

100 g of impure NaCl contain pure $NaCl = 95$

1 g of impure NaCl contains pure $NaCl = \dfrac{95}{100}$

250 g of impure NaCl contains pure $NaCl = \dfrac{95}{100} \times 250$

$= 237.5$ g.

Step IV

Find the amount of pure Na_2SO_4 obtained from 237.5 of pure NaCl.

From the equation we see that

2×58.5 g of NaCl gives $Na_2SO_4 = 142$ g

1 g of NaCl gives $Na_2SO_4 = \dfrac{142}{117.0}$ g

237.5 g of NaCl will give $Na_2SO_4 = \dfrac{142}{117.0} \times 237.5$ g

$= 288.2$ g.

Step V

Find the amount of 83.4% Na_2SO_4 frpm 288.2 g pure Na_2SO_4.

83.4% Na_2SO_4 means that if the impure sample of Na_2SO_4 is 100, pure sodium sulphate present in it = 83.4 g.

or we can say that pure 83.4 g Na_2SO_4 gives impure

$= 100$ g

or pure 1 g Na_2SO_4 gives impure $= \dfrac{100}{83.4}$ g

or pure 288.2 g Na_2SO_4 gives impure $= \dfrac{100 \times 288.2}{83.4}$ g

$= 345.4$ g.

Example 4. *Air contains 21% oxygen by weight. What weight of air is required to burn 200 g of coal which contains only 80% combustible material ?*

Step I

Write the chemical equation involved.

Coal is carbon ; on combustion in oxygen it gives CO_2. Thus the equation is

$$C + O_2 \longrightarrow CO_2$$

Step II

Find the weight of combustible matter in 200 g of coal.

Since coal contains only 80% combustible matter this means that 100 g of coal has carbon = 80 g

∴ 200 g of coal will have carbon = $\dfrac{80}{100} \times 200$ g

= 160 g.

Step III

Find the weight of oxygen needed for combustion.

Now from the equation we see that

12 g of carbon requires oxygen = 32 g

1 g of carbon requires oxygen = $\dfrac{32}{12}$ g

160 g of carbon requires oxygen = $\dfrac{32}{12} \times 160$ g

= 426·66 g.

Step IV

Find the weight of air.

Now air contains 21% oxygen by weight

i.e., 21 g oxygen makes air = 100 g

1 g of oxygen makes air = $\dfrac{100}{21}$ g

and 426·66 g of oxygen makes air = $\dfrac{100}{21} \times 426 \cdot 66$ g

= 2031·7 g of air.

Example 5. *A mixture of the carbonates of calcium and magnesium weighing 1·42 g was strongly heated until no further loss was produced. The weight of the residue obtained was 0·76 g. Find the percentage composition of the mixture.*

Step I

Write the equations involved.

$$CaCO_3 \xrightarrow{heat} CaO + CO_2$$

$$MaCO_3 \xrightarrow{heat} MgO + CO_2$$

Step II

Find the molecular weights of the chemical compounds involved in the problem.

Mol. wt. of $CaCO_3 = 40+12+48 = 100$

Mol. wt. of $CaO = 40+16 = 56$

Mol. wt. of $MgCO_3 = 24+12+48 = 84$

Mol. wt. of $MgO = 24+16 = 40$.

Step III

Find the theoretical weight of the residue CaO and MgO obtained from $CaCO_3$ and $MgCO_3$ on heating.

Let the amount of $CaCO_3$ in the mixture $= x$ g

∴ the amount of $MgCO_3$ in the mixture $= 1.42-x$ g

From the equation we find that

100 g of $CaCO_3$ gives $CaO = 56$ g

1 g of $CaCO_3$ gives $CaO = \dfrac{56}{100}$ g

x g of $CaCO_3$ gives $CaO = \dfrac{56}{100} \times x$ g

$\qquad\qquad\qquad\qquad\quad = 0.56\, x$ g

Also

84 g of $MgCO_3$ gives $MgO = 40$ g

1 g of $MgCO_3$ gives $MgO = \dfrac{40}{84}$ g

$(1.42-x)$ g of $MgCO_3$ gives $MgO = \dfrac{40}{84}(1.42-x)$ g

$\qquad\qquad\qquad\qquad\qquad\quad = \dfrac{10}{21}(1.42-x)$ g

Total theoretical weight of the two oxides

$$= 0.56\,x + \dfrac{10}{21}(1.42-x) \text{ g}.$$

Step IV

Find the value of x.

The actual weight of the residue left after the ignition of 1.42 g mixture is $= 0.76$ g

(The rest has escaped in the form of CO_2)

Thus,

$$0.56\, x + \frac{10}{21}(1.42 - x) = 0.76$$

Simplifying the above equation,

$$x = 1$$

∴ The amount of $CaCO_3$ in the mixture $= 1$ g

and the amount of $MgCO_3$ in the mixture $= 0.42$ g

STEP V

Find the percent of each component.

$$\text{Percentage of } CaCO_3 = \frac{1}{1.42} \times 100$$

$$= 70.42$$

and the percentage of $MgCO_3 = 100 - 70.42$

$$= 29.58.$$

Example 6. *1.2 g of a sample of common salt when treated with excess of silver nitrate solution gave 2.869 g of silver chloride. Calculate the percentage purity of common salt.*

STEP I

Write the chemical equation involved.

$$NaCl + AgNO_3 \longrightarrow NaNO_3 + AgCl$$

STEP II

Find the molecular weights of the substances involved in the problem.

Molecular weight of $NaCl = 23 + 35.5 = 58.5$

Molecular weight of $AgCl = 108 + 35.5 = 143.5$ g

STEP III

Find the weight of NaCl from the weight of AgCl.

From the equation we see that

143·5 g of AgCl is obtained from NaCl = 58·5 g

1 g of AgCl is obtained from NaCl = $\dfrac{58·5}{143·5}$ g

2·869 g of AgCl is obtained from NaCl

$$= \dfrac{58·5}{143·5} \times 2·869 = 1·169 \text{ g}$$

Step IV

Find the percentage purity of NaCl.

Actually we took 1·2 g of common salt but the reaction has established that only 1·169 g is pure NaCl in 1·2 g common salt; the rest may be an impurity.

Thus 1·2 g common salt contains pure NaCl = 1·169 g

∴ 1 g common salt contains pure NaCl = $\dfrac{1·169}{1·2}$ g

and 100 g common salt contains pure NaCl

$$= \dfrac{1·169}{1·2} \times 100 = 97·41 \text{ g}$$

Hence percentage purity of NaCl = 97·41.

Example 7. *A certain weight of sodium iodide and sodium chloride mixture when treated with sulphuric acid was found to give the same weight of sodium sulphate. Calculate the percentage composition of the mixture.*

Step I

Write the equations involved.

$$2NaCl + H_2SO_4 \longrightarrow Na_2SO_4 + 2HCl$$
$$2NaI + H_2SO_4 \longrightarrow Na_2SO_4 + 2HI$$

Step II

Find the molecular weights of the compounds involved in the problem.

Mol. wt. of NaCl = 23 + 35·5 = 58·5

Mol. wt. of NaI = 23 + 127 = 150

Mol. wt. of Na_2SO_4 = 142.

Step III

Find the theoretical weight of Na_2SO_4 obtained from NaCl and NaI.

Let the weight of the mixture = 1 g
Let the weight of NaCl = x g
and the weight of NaI = $(1-x)$ g

From the equation we see that

2×58.5 g of NaCl gives Na_2SO_4 = 142 g

1 g of NaCl gives $Na_2SO_4 = \dfrac{142}{117}$ g

x g of NaCl gives $Na_2SO_4 = \dfrac{142x}{117}$ g

Similarly,

300 g of NaI gives Na_2SO_4 = 142 g

1 g of NaI gives $Na_2SO_4 = \dfrac{142}{300}$ g

$(1-x)$ g of NaI gives $Na_2SO_4 = \dfrac{142}{300} \times (1-x)$ g

Total theoretical weight of Na_2SO_4

$$= \dfrac{142x}{117} + \dfrac{142}{300}(1-x) \text{ g}$$

Step IV

Find the value of x.

Now actual amount of Na_2SO_4 formed

= 1 g (same weight)

∴ $\dfrac{142x}{117} + \dfrac{142}{300} \times (1-x) = 1$.

or = 0.7114

∴ Wt. of NaCl = 0.7114 g

and the wt. of NaI = 1 − 0.7114
= 0.2886 g

Step V

Find the per cent of each component.

Now 1 g of the mixture contains NaCl = 0·7114 g

∴ 100 g of the mixture contains NaCl = 0·7114 × 100 g
$$= 71.14 \text{ g}$$

and the %age of NaI = 100 − 71·14
$$= 28.86$$

Example 8. *1·240 g of pure $CaCO_3$ was completely neutralized by 48·4 ml of HCl. Find the strength of HCl solution by volume when the specific gravity of HCl solution is 1·12 g/ml.*

Step I

Write the equation involved.

$$CaCO_3 + 2HCl \longrightarrow CaCl_2 + H_2O + CO_2$$

Step II

Find the mol. wt. of the compounds involved in the problem.

Mol. wt. of $CaCO_3 = 40 + 12 + 48 = 100$

Mol. wt. of $HCl = 1 + 35.5 = 36.5$

Step III

Find the weight of HCl required to neutralize 1·24 g of $CaCO_3$

From the equation we see that

100 g of $CaCO_3$ required for complete neutralization
$$HCl = 2 \times 36.5 = 73 \text{ g}$$

1 g $CaCO_3$ will need $HCl = \dfrac{73}{100}$ g

and 1·24 g of $CaCO_3$ will need $HCl = \dfrac{73}{100} \times 1.24$ g
$$= 0.9052 \text{ g.}$$

Step IV

Find the strength of HCl solution.

Actually we are given 48·4 ml of HCl solution.

Problems Based on Equations 197

The above calculations show that only 0·9052 g HCl is present.

Since the specific gravity is 1·12 g/ml, it means that 1 ml of this solution weighs 1·12 g

∴ 48·4 ml of HCl solution will weigh $= \dfrac{1\cdot12}{1} \times 48\cdot5$

$= 54\cdot208$ g

Thus 54·208 g solution of HCl contains pure HCl
$= 0\cdot9052$ g

∴ 100 g solution of HCl contains pure HCl

$= \dfrac{0\cdot9052}{54\cdot208} \times 100$

$= 1\cdot67$ g.

Thus strength of the given HCl solution is **1·67%** HCl.

REACTIONS IN SUCCESSION

Example 9. *How much of potassium chlorate is needed to get enough oxygen for completely burning 4·0 g of carbon*

STEP I

Write the equations involved.

$$2KClO_3 \xrightarrow{\text{heat}} 2KCl + 3O_2$$

$$\dfrac{[C + O_2 \longrightarrow CO_2] \times 3}{2KClO_3 + 3C \longrightarrow 2KCl + 3CO_2}$$

STEP II

Find the molecular or atomic weight of the compound involved.

Molecular weight of $KClO_3 = 39 + 35\cdot5 + 48 = 122\cdot5$

at. wt. of $C = 12$.

STEP III

Find the weight of $KClO_3$.

From the equation we find that

2×122.5 g are needed for the complete combustion of 3×12 g of carbon. Thus,

36 g of carbon needs $KClO_3 = 245.0$ g

1 g of carbon needs $KClO_3 = \dfrac{245}{36}$ g

4 g of carbon needs $KClO_3 = \dfrac{245}{36} \times 4$

$= 27.22$ g

Thus the amount of $KClO_3$ needed for the complete combustion of 4.0 g of carbon $= 27.22$ g.

Example 10. *16.0 g of MnO_2 are heated with excess of hydrochloric acid and the gas evolved is passed into a solution of potassium iodide. Calculate the weight of iodine that is liberated.*

STEP I

Write the equations involved.

$$MnO_2 + 4HCl \longrightarrow MnCl_2 + 2H_2O + Cl_2$$
$$2KI + Cl_2 \longrightarrow 2KCl + I_2$$

The resultant equation obtained after cancellation is

$$MnO_2 + 4HCl + 2KI \longrightarrow MnCl_2 + 2KCl + 2H_2O + I_2$$

STEP II

Find the molecular weight of the compounds involved in the problem.

Molecular weight of $MnO_2 = 55 + 32 = 87$

Molecular weight of $I_2 = 2 \times 127 = 254$.

STEP III

From the resultant equation we see that

87 g of MnO_2 liberates iodine $= 254$ g

1 g of MnO_2 will liberate iodine $= \dfrac{254}{87}$ g

16 g of MnO_2 will liberate iodine $= \dfrac{254}{87} \times 16$

$= 46.71$ g.

END-OF-CHAPTER PROBLEMS

1. Calculate the weight of iron oxide which will be obtained by heating 6·25 g of iron according to the equation

$$3Fe + 2O_2 \rightarrow Fe_3O_4$$

[**Ans.** 8·632 g]

2. What weight of water should be decomposed by electrolysis to get 40·0 g of hydrogen

$$(2H_2O \rightleftharpoons 2H_2 + O_2)$$

[**Ans.** 360 g]

3. 100 g of marble chips are dropped into 500 g of a solution of HCl containing one-tenth of its weight of pure acid. How much of the chips will remain undissolved and what weight of anhydrous $CaCl_2$ could be obtained from the solution?

[**Ans.** Marble chips left undissolved = 31·6 g
Wt. of $CaCl_2$ = 76·03 g]

4. What weight of sulphur is needed to react with 6·50 g of copper in the preparation of CuS? What weight of CuS will be produced? [**Ans.** 3·30 g of S, 9·76 g of CuS]

5. What weight of iron would be produced by the reduction of 1 kg of ferric oxide? [**Ans.** 0·70 kg]

6. What weight of phosphorus is required for the preparation of 11·5 g of phosphorus pentoxide? [**Ans.** Phosphorus = 5·02 g]

7. What weight of zinc sulphate is obtained by the action of 4·4 g of zinc with dilute H_2SO_4 solution?

[**Ans.** 10·8 g of zinc sulphate]

8. Air contains 21% oxygen by weight. What weight of air is required for the complete combustion of 24 g of ethane (C_2H_6).

$$2C_2H_6 + 7O_2 \rightarrow 4CO_2 + 6H_2O$$

[**Ans.** 426·66 g of air]

9. What weight of $CaCO_3$ must be heated to get 70 g of lime (CaO)? What weight of CO_2 will be produced at the same time?

[**Ans.** 125 g of $CaCO_3$, 55 g of CO_2]

10. A sample of rock containing 40% FeS was burnt to give SO_2. What weight of SO_2 will be obtained by burning 500 g of such a rock? [**Ans.** 145·55 g of SO_2]

11. 5·0 g of iceland spar ($CaCO_3$) were added to 7·5 g of dilute HCl. After the reaction was over it was found that 0·50 g of iceland spar was left undissolved. Calculate the percentage purity of the acid

[**Ans.** 43·8% HCl]

12. How many grams of hydrogen are obtained from 50 g of CaH_2 by treatment with water,

$$CaH_2 + 2H_2O \rightarrow Ca(OH)_2 + 2H_2$$

[**Ans.** 4 g]

13. A 1·425 g sample of an organic compound was burnt completely. 1·771 g of CO_2 and 0·725 g of water was formed. The compound contains only carbon and hydrogen. Find the empirical formula of the compound. [**Ans.** $C_2H_6O_4$]

14. Which is cheaper, 40% HCl at the rate of 65 paise per kilogram or 80% sulphuric acid at the rate of 65 paise per Kilogram needed to completely neutralize 7 kg of KOH ? [**Ans.** Sulphuric acid]

15. A silver coin weighing 2·91 g was dissolved in dil. HNO_3. When all silver had dissolved a solution of hydrochloric acid was added. 3·60 g of AgCl was obtained. Determine the percentage of silver in the coin. [**Ans.** 93·1%]

16. How much 50% H_2SO_4 is required to decompose 50 gm. of chalk, and how much of calcium sulphate is formed.
[**Ans.** 98 gm, 68 gm]

17. Calculate the weights of $ZnCO_3$ and HCl required to produce 22·4 litres of CO_2 at N.T.P. What weight of $ZnCl_2$ will be produced in the reaction ?

(At wts, Zn=65 ; C=12 ; O=16 ; H=1 ; Cl=35·5)
(Punjab Board Hr. Sec., 1978)

[**Ans.** $ZnCO_3$=125 g
HCl=73 g
$ZnCl_2$=136 g]

18. What volume of CO_2 gas will be evolved by the complete decomposition of 6·25 g of $CaCO_3$ on heating ?
(Punjabi Pre-Univ., 1978)
[**Ans.** 1400 ml]

19. Limonite ($Fe_2O_3 \cdot H_2O$) contains 47·136% of the ore and the rest is silicon dioxide present as impurity. 454 kg of limonite is treated with excess of coal and calculated amount of lime stone for the extraction of iron :

$$Fe_2O_3 + 3C \rightarrow 2Fe + 3CO$$

Find out :

(i) How many moles of iron are produced ?

(ii) How many moles of water are given out ?

(iii) How many moles of slag are formed ?

(iv) What volume of carbon monoxide is formed during the reduction of the ore at NTP ?

(At. Wts. H = 1, C = 12, Si = 28, Ca = 40, Fe = 56, N = 14)
(Punjab Pre-Univ., 1981)

[**Ans.** (i) moles of iron = 2000
(ii) moles of water = 3000
(iii) moles of slag = 4000
(iv) volume of Co = 67200 litres]

Problems Based on Equations

20. With the help of chemical equation given below calculate the amounts of silver nitrate and sodium chloride required to produce 2·87 g of silver chloride.

$$AgNO_3 + NaCl \rightarrow AgCl + NaNO_3$$

(At. Wts. Ag = 108 ; N = 14 ; O = 16 ; Na = 23 ; Cl = 35·5)

(Punjabi Pre-Univ., 1981)

[**Ans.** $AgNO_3$ = 3·4 g, NaCl = 1·17 g]

Mixture

21. Current market prices of Al, Zn, and Fe Scraps per Kg. are Rs. 30/-, Rs. 24/-, and Rs. 4/- respectively. If hydrogen is to be prepared by the reaction of one of the metals with dilute H_2SO_4, which would be the cheapest and which would be expansive metal ? (At. wts : Al=27, Zn=65, Fe=56). *(Punjabi Pre-Univ., 1982)*

[**Ans.** Cheapest would be Fe ; Expensive would be Zn]

22. 1·48 g of a mixture of calcium and magnesium carbonates gave on ignition 0·96 g of a residue. What is the percentage composition of the mixture ? [**Ans.** Percentage of $CaCO_3$=54·48 ;

„   „ $MgCO_3$=45·52]

23. 10·0 g of a mixture of anhydrous nitrates of two metals A and B were heated to a constant weight and gave 5·531 g of mixture of the corresponding oxides. The equivalent weights of A and B are 103·6 and 31·8 respectively, What was the percentage of A in the mixture ? [**Ans.** 51·62%]

24. One gram of a mixture of potassium and sodium chlorides on treatment with excess of $AgNO_3$ gave two grams of silver chloride. What was the proportion of the two salts in the original mixture ?

[**Ans.** 6·14 ; 1]

25. A quantity of a mixture of KI and NaCl changed into normal sulphates is found to weigh the same as the original salts. Find the composition of the mixture. [**Ans.** KI=30·98% ;

NaCl=69·02%]

26. 4·5 g of a mixture of sodium bicarbonate and sodium carbonate was heated strongly to give a constant weight of the residue. The weight of the residue was 3·105 g. Calculate the percentage composition of the mixture. [**Ans.** $NaHCO_3$=85% ;

Na_2CO_3=15%]

27. 2·432 g of mixture of carbonates of calcium and barium was ignited till constant weight. The residue was found to weigh 1·6352 g. Calculate the percentage composition of the mixture.

[**Ans.** $CaCO_3$=48·11 % ;

$BaCO_3$=51·89 %]

28 3·68 gm. of a mixture of calcium carbonate and magnesium carbonate on heating to a constant weight produced 1.92 gm of residue. Find the percentage composition of the given mixture. (Atomic masses : H=1, C=12, O=16, Ca=40, Mg=24.)

(Punjab Pre-Univ, 1988)

[**Ans.** $CaCO_3$=36.14% ; $MgCO_3$=63.86%]

Percentage Purity

29. 35 g of sample of impure zinc reacts with 90·3 ml of hydrochloric acid which has specific gravity 1·18 and contains 30% HCl by weight. What is the percentage purity of the zinc metal ? Assume that the impurity does not react with HCl.   [**Ans.** 96% zinc]

30. When copper is heated with sulphur, Cu_2S is formed. 110 g of an impure sample of copper metal gave 125 g of Cu_2S. Find the percentage purity of copper in the given sample. Assuming that the impurity does not react with sulphur.   [**Ans.** 90·90% Cu]

31. 5 g of an impure sample of sodium carbonate was completely neutralized by 8·84 ml of 35% HCl (by weight) and specific gravity 1·18. Calculate the percentage purity of Na_2CO_3 in the given sample.
[**Ans.** 73·1%]

32. 10·5 g of an impure sample of $CaCO_3$ was heated strongly and 3·5 of CO_2 g was collected. Find the percentage purity of the given sample of $CaCO_3$.   [**Ans.** 75·7%]

33. 0·2 g of a sample of common salt solution when treated with excess of silver nitrate solution gave 0·28 g of AgCl. Calculate the percentage of NaCl in the com on salt.   [**Ans.** 47%]

34. 0·3 g of a sample of common salt when treated with silver nitrate solution gave 0·685 g of silver chloride. Calculate the percentage purity of common salt.   (*Rajasthan Pre-Univ., 1982*)
[**Ans.** 93·1%]

35. 15 g of an impure sample of sodium sulphate dissolved in water was treated with excess of barium chloride solution. 1·74 g of $BaSO_4$ was obtained as precipitate. Calculate the percentage purity of the sample.

[Na=23, S=32, O=16, Ba=137]
(*Haryana Board Hr. Sec., 1979*)
[**Ans.** 70·69%]

Reactions in Succession

36. How much potassium nitrate is needed to get oxygen enough to completely react with 5 g of hydrogen ?
[**Ans.** 252·5 g of KNO_3]

37. 16·0 g of MnO_2 is heated with excess of HCl and the gas is passed into KI solution. Calculate the weight of iodine that is liberated ?   [**Ans.** 16·71 g]

38. Calculate the weight of iron sulphide required to produce sufficient H_2S which would precipitate 1·127 g of zinc solution.
[**Ans.** 0·616 g FeS]

39. 50 g caustic soda were completely converted into sodium chlorate and sodium chloride by the action of chlorine. What weight of manganese dioxide and what volume of HCl solution (containing 300 g acid per litre) were used for the production of the necessary amount of chlorine ?   [**Ans.** MnO_2 ; 54·375 g ;

HCl=304·167 ml]

Problems Based on Equations

40. How much MnO_2 and how much hydrochloric acid containing 33% HCl would be required to furnish chlorine necessary to convert 56 g of caustic potash into chlorate and chloride?

(*Ranchi Pre-Univ.*, 1971)

[**Ans.** 43·5 g. 221·2 g]

41. What weight of zinc and sulphuric acid would be required to give enough hydrogen to reduce 32 g of ferric oxide.

(*Punjab H.S.* 1972)

[**Ans.** 39 g, 58·8 g]

42. 17·4 g of manganese dioxide is heated with excess of HCl and the gas evolved is passed into a solution of potassium iodide. Calculate the weight of potassium iodide decomposed and iodine liberated.

(*Punjab H.S.* 1973)

[**Ans.** 66·4 g ; 50·8 g]

43. Calculate the weight of pure manganese dioxide and hydrochloric acid of 80% purity required to get enough chlorine to convert 12 g of sodium hydroxide into sodium chlorate.
(H = 1, O = 16, Na = 23, Cl = 35·5, Mn = 55)

(*Kurukshetra Pre-Univ.*, 1981)

[**Ans.** MnO_2 = 13·04 g

HCl = 26·6 g]

Miscellaneous

44. Iodine is prepared according to the following reaction

$$2NaIO_3 + 5NaHSO_3 \longrightarrow 3NaHSO_4 + 2Na_2SO_4 + H_2O + I_2$$

How much $NaIO_3$ and $NaHSO_3$ will be required to produce 1 kg. of iodine?

[**Ans.** $NaIO_3$ = 1·56 kg ;

$NaHSO_3$ = 2·05 kg]

45. A sample of impure Cu_2O contains 66·6% copper. What is the percentage of pure Cu_2O in the sample?

[**Ans.** 75% Cu_2O]

46. What weight of ammonium sulphate is required to give the same weight of ammonia as 35 g of ammonium chloride?

[**Ans.** 43·225 g]

47. What weight of sulphuric acid can be obtained from sulphur dioxide formed by the combustion of 25 g of sulphur?

[**Ans.** 76·5 g]

48. A solution contains 1·5 g of sodium sulphate. What weight of $BaCl_2$ is required to precipitate the sulphate completely as barium sulphate? What weight of $BaSO_4$ will be obtained?

[**Ans.** 2·199 g of $BaCl_2$;

2·463 g of $BaSO_4$]

49. What weight of potassium dichromate is required to oxidise 15 g of ferrous sulphate?

[**Ans.** 4·842 g]

50. 10 g of a commercial sample of $CaCO_3$ gave on treatment with excess HCl, 3·3 g of CO_2. Calbulate the percentage purity of the sample.   (*Poona Pre-Univ.*, 1976)

[**Ans.** 75%]

51. 2·5 g of impure potassium chlorate gave enough oxygen for complete combustion of 1 g of carbon monoxide. Calculate the percentage purity of potassium chlorate.

[At. wts. ; K=39, Cl=35·5, O=16, C=12]
(*Madurai Pre-Univ.*, 1975)

[**Ans.** 58·33]

52. 1·84 g of a mixture of calcium carbonate and magnesium carbonate were heated strongly till no further loss in weight. The residue weighed 0·95 g. Find the percentage composition of the given mixture.   (*Kurukshetra Pre-Univ.*, 1976)

[**Ans.** $CaCO_3$=54·35%)
$MgCO_3$=45·65%]

53. What weight of magnesium and sulphuric acid would be required to produce enough hydrogen at N.T.P to reduce completely 15 g of copper oxide (CuO) to copper.

[H=1, O=16, Mg=24, Cu=63·5, S=32]
(*Haryana Board Hr. Sec.*, 1979)

[**Ans.** Mg=4·524 g
H_2SO_4=18·473 g]

14

Problems Based on Equations (Contd.)

TYPE II—(Weight-Volume Relationship)

Here we come across problems wherein some or all of the reactants and products are gaseous and we have to measure their volumes instead of weights. In all such types of problems we make use of the principle of gram molecular volume according to which **22·4 litres of any gas at N.T.P. possesses a weight equal to its molecular weight in grams (G.M.W.)** It should be noted that the volumes of the reactants or the products (if gaseous) represented in the chemical equation are taken to be at S.T.P.

For instance we have a reaction

$$2SO_2 + O_2 \rightleftharpoons 2SO_3$$

Here the 2 volumes of sulphur dioxide (gas) and one volume of oxygen (gas) means $2 \times 22·4$ litres of SO_2 and $1 \times 22·4$ litres of oxygen are reacting between them at S.T.P. If the volumes of the gaseous reactants or the products are desired at any other temperature and pressure, the same can be calculated by applying the gas equation.

Example 1. *What volume of carbon dioxide will be produced by heating 250 g of calcium carbonate?*

STEP I

Write the chemical equation involved.

$$CaCO_3 \longrightarrow CaO + CO_2$$

Step II

Find the volume of CO_2.

From the equation we find that 100 g (G.M.W.) of $CaCO_3$ give one molecule or 22·4 litres of CO_2.

Thus 100 g of $CaCO_3$ gives $CO_2 = 22·4$ litres

$$1 \text{ g of } CaCO_3 \text{ gives } CO_2 = \frac{22·4}{100}$$

$$250 \text{ g of } CaCO_3 \text{ gives } CO_2 = \frac{22·4}{100} \times 250 = 56 \text{ litres.}$$

Hence 250 g of $CaCO_3$ will give 56 litres of CO_2 at S.T.P.

Example 2. *What volume of hydrogen at S.T.P. will be released by the action of 17·5 g of magnesium on dilute sulphuric acid ?*

Step I

Write the chemical equation involved.

$$Mg + H_2SO_4 \longrightarrow MgSO_4 + H_2$$

Step II

Find the volume of hydrogen evolved at S.T.P.

From the equation we see that one gram atom or 24·0 g of magnesium produce 1 molecule or 22·4 litres of hydrogen at S.T.P.

Thus,

24·0 g of Mg produces hydrogen = 22·4 litres

$$1 \text{ g of Mg produces hydrogen} = \frac{22·4}{24}$$

$$17·5 \text{ g of Mg produces hydrogen} = \frac{22·4}{24} \times 17·5$$
$$= 16·3 \text{ litres.}$$

Hence 17·5 of magnesium metal will produce **16·3** litres of H_2 at S.T.P.

Example 3. *When calcium carbonate is heated to 1000°C it decomposes with the formation of carbon dioxide and calcium* oxide. *What volume of the gas (CO_2) will be produced by one pound of $CaCO_3$ at the temperature of the reaction ($1000°C$) and 700 mm pressure ?*

STEP I

Write the chemical equation involved.

$$CaCO_3 \longrightarrow CaO + CO_2$$

STEP II

Find the volume of CO_2 at S.T.P.

From the equation we see that

100 g of $CaCO_3$ gives $CO_2 = 22.4$ litres

1 g of $CaCO_3$ gives $CO_2 = \dfrac{22 \cdot 4}{100}$ litres

1 lb (or 4·53·59 g) of $CaCO_3$ will give CO_2

$$= \dfrac{22 \cdot 4}{100} \times 453 \cdot 59$$

$$= 101 \cdot 664 \text{ litres.}$$

STEP III

Find the volume of CO_2 at $1000°C$ and 700 mm.

Applying the gas equation

$$\dfrac{P_1 V_1}{T_1} = \dfrac{P_2 V_2}{T_2}$$

when
$P_1 = 760$ mm
$V_1 = 101 \cdot 664$ litres
$T_1 = 273°$ Abs.
$P_2 = 700$ mm
$V_2 = ?$
$T_2 = 1000 + 273 = 1273°$ Abs.

We have

$$\dfrac{760 \times 101 \cdot 664}{273} = \dfrac{700 \times V_2}{1273}$$

$$V_2 = \dfrac{760 \times 101 \cdot 664}{273} \times \dfrac{1273}{700}$$

$$= 514 \cdot 6 \text{ litres.}$$

Hence the volume of CO_2 produced by 1 lb of $CaCO_3$ at 1000°C and 700 mm pressure is 514·6 litres.

Example 4. *The electrolysis of brine takes place according to the equation*

$$2NaCl + 2H_2O \rightleftharpoons H_2 + Cl_2 + 2NaOH$$

What volume of hydrogen is liberated at S.T.P. by the electrolysis of 1 kg of NaCl ?

The equation is already given.

The equation tells that $2 \times 58·5$ g of NaCl gives 22·4 litres of hydrogen gas at S.T.P.

So, 1 kg or (1000 g) NaCl will give $\dfrac{22·4}{117} \times 1000$ or 191·45 litres of H_2 at S.T.P.

Example 5. *What weight of Barium peroxide is required for the liberation of 200 litres of oxygen to be measured at 15°C and 710 mm pressure ?*

STEP I

Write the chemical equation involved.

$$2BaO_2 \longrightarrow 2BaO + O_2$$

STEP II

Find the vol. of oxygen at S.T.P.

Here
$V_1 = 200$ litres
$T_1 = 273 + 15 = 288°$ Abs.
$P_1 = 710$ mm
$P_2 = 760$ mm
$V_2 = \ ?$
$T_2 = 273°$ Abs.

Applying the gas equation

$$\dfrac{P_1 V_1}{T_1} = \dfrac{P_2 V_2}{T_2}$$

Problems Based on Equations (Contd.)

and substituting the values we have

$$\frac{710 \times 200}{288} = \frac{760 \times V_2}{273}$$

or $\qquad V_2 = \dfrac{710 \times 200 \times 273}{288 \times 760}$ 177·1 litres

Step III

Find the weight of BaO_2.

The equation tells that

22·4 litres of oxygen is obtained from $BaO_2 = 2 \times 169$ g

1 litre of oxygen is obtained from $BaO_2 = \dfrac{2 \times 169}{22 \cdot 4}$ g

177·1 litres of oxygen is obtained from BaO_2

$$= \frac{2 \times 169}{22 \cdot 4} \times 177 \cdot 1 = 2672 \cdot 3 \text{ g.}$$

Thus weight of BaO_2 required = 2672·3 g

Example 6. *A piece of zinc when placed in 50 ml of a solution of HCl gave 304 ml of hydrgen measured at 15°C and 740 mm pressure. Calculate the normality of the solution of hydrochoric acid.*

Step I

Write the chemical equation involved.

$$Zn + 2HCl \longrightarrow ZnCl_2 + H_2.$$

Step II

Find the volume of hydrogen at N·T·P.

$P_1 = 740$ mm	$P_2 = 760$ mm
$V_1 = 304$ ml	$V_2 = ?$
$T_1 = 273 + 15 = 288°$ Abs	$T_2 = 273°$ Abs.

Applying the gas equation

$$\frac{P_1 V_1}{T_1} = \frac{P_2 V_2}{T_2}$$

14—NPC

and substituting the values we have

$$\frac{740 \times 304}{288} = \frac{760 \times V_2}{273}$$

or $\quad V_2 = \dfrac{740 \times 304 \times 273}{288 \times 760} = 280 \cdot 6$ ml.

STEP III

Find the weight of HCl.

From the equation we find that

22400 ml of hydrogen at N.T.P. is obtained from HCl
= 73 g

1 ml of hydrogen at N.T.P. is obtained from

$$HCl = \frac{73}{22400} \text{ g}$$

280·6 ml of hydrogen at N.T.P. is obtained from HCl

$$= \frac{73}{22400} \times 280 \cdot 6 = 0 \cdot 9145 \text{ g HCl.}$$

STEP IV

Find the strength and normality of HCl.

Now 50 ml of HCl contains pure HCl = 0·9145 g

∴ 1 ml of HCl contains pure HCl = $\dfrac{0 \cdot 9145}{50}$

and 1000 ml of HCl contains pure HCl

$$= \frac{0 \cdot 9145}{50} \times 1000 = 18 \cdot 290 \text{ g}$$

$$\text{Normality} = \frac{\text{Strength}}{\text{Eq. wt.}} = \frac{18 \cdot 29}{36 \cdot 5} = 0 \cdot 5 \text{ N}$$

Thus, the normality of given HCl solution is 0·5 N.

Example 7. *2·5 g of potassium chlorate on heating gave 580 ml of oxygen at 17°C and 750 mm pressure. Assuming the dicomposition to be complete, calculate the percentage purity of the sample.*

STEP I

Write the chemical equation involved,

$$2KClO_3 \longrightarrow 2KCl + 3O_2$$

Step II

Find the volume of oxygen at S.T.P.

$P_1 = 750$ mm $P_2 = 760$
$V_1 = 580$ ml $V_2 = ?$
$T_1 = 273 + 17 = 290°$ Abs. $T_2 = 273°$ Abs

Applying the gas equation and substituting the values we get

$$\frac{750 \times 580}{290} = \frac{760 \times V_2}{273}$$

or

$$V_2 = \frac{750 \times 580 \times 273}{290 \times 760} = 538 \cdot 9 \text{ ml}$$

Step III

Find the weight of $KClO_3$.

From the equation it is seen that

3×22400 ml of oxygen are obtained from $KClO_3$
$= 245 \cdot 0$ g

1 ml of oxygen are obtained from $KClO_3 = \dfrac{245}{3 \times 22400}$

$538 \cdot 9$ ml of oxygen are obtained from $KClO_3$

$$= \frac{245}{3 \times 22400} \times 538 \cdot 9 = 1 \cdot 96 \text{ g}.$$

Step IV

Find the percentage purity of $KClO_3$.

The total weight of the impure sample of $KClO_3 = 2 \cdot 5$ g and the quantity of pure $KClO_3 = 1 \cdot 96$ g

Thus $2 \cdot 5$ g impure sample has pure $KClO_3 = 1 \cdot 96$ g

1 g impure sample has pure $KClO_3 = \dfrac{1 \cdot 96}{2 \cdot 5}$ g

100 g impure sample has pure $KClO_3$

$$= \frac{1 \cdot 96}{2 \cdot 5} \times 100 \text{ g}$$
$$= 78 \cdot 4 \text{ g}$$

Hence percentage purity of $KClO_3 = 78.4$.

Example 8. *One gram of mixture of calcium carbonate and magnesium carbonate gave 240 ml of CO_2 at S.T.P. Calculate the composition of the mixture.*

STEP I

Write the chemical equations involved.

$$CaCO_3 \longrightarrow CaO + CO_2$$
$$MgCO_3 \longrightarrow MgO + CO_2.$$

STEP II

Find the total theoretical volume of CO_2 obtained from the mixture.

Let the wt. of $CaCO_3$ present in 1 g of mixture

$$= x \text{ g}$$

so that the wt. of $MgCO_3$ present in 1 g of the mixture

$$= (1-x) \text{ g}.$$

From the equation we see that

100 g of $CaCO_3$ gives $CO_2 = 22.4$ litres

1 g of $CaCO_3$ gives $CO_2 = \dfrac{22.4}{100}$

and x g of $CaCO_3$ gives $CO_2 = \dfrac{22.4}{100} \times x$ litres.

Similarly,

84 g of $MgCO_3$ given $CO_2 = 22.4$ litres

1 g of $MgCO_3$ gives $CO_2 = \dfrac{22.4}{84}$ litres

$(1-x)$ g of $MgCO_3$ given $CO_2 = \dfrac{22.4}{84} \times (1-x)$ litres

Total volume of $CO_2 = \dfrac{22.4\ x}{100} + \dfrac{22.4}{84}(1-x)$ litres

STEP III

Find the volume of x_2

Now actual volume of CO_2 produced $= 240$ ml

Problems Based on Equations (Cond.)

$$= \frac{240}{1000} \text{ litre} = 0.240 \text{ litre.}$$

Thus $\quad \dfrac{22 \cdot 4x}{100} + \dfrac{22 \cdot (41-x)}{84} = 0.240$

or $\quad x = 0.6253$

Thus the weight of $CaCO_3 = 0.6253$ g
and the weight of $MgCO_3 = 1 - 0.6253 = 0.3747$ g

Thus percentage of $CaCO_3 = \dfrac{0.6253}{1} \times 100 = 62.30$

and percentage of $MgCO_3 = 100 - 62.30 = 37.40$.

Example 9. *1 g of a mixture of anhydrous sodium carbonate and sodium bicarbonate when heated gave 127.5 ml of CO_2 at 27°C and 700 mm pressure. Find the percentage composition of the mixture.*

Step I

Write the chemical equation involved.

Alkali carbonates do, not decompose on heating. Bicarbonates however decompose to give carbonates and CO_2. Thus

$$2NaHCO_3 \longrightarrow Na_2CO_3 + CO_2 + H_2O$$

Step II

Find the volume of CO_2 at S.T.P.

Now $\quad P_1 = 700$ mm $\qquad P_2 = 760$ mm
$\quad V_1 = 127.5$ ml $\qquad V_2 = ?$
$\quad T_1 = 300°$ Abs. $\qquad T_2 = 273°$ Abs.

Substituting the values in the equation

$$\frac{P_1 V_1}{T_1} = \frac{P_2 V_2}{T_2}$$

we have $\quad \dfrac{700 \times 127.5}{300} = \dfrac{760 \times V_2}{273}$

or $\quad V_2 = \dfrac{700 + 127.5 \times 273}{300 \times 760} = 106.9$ ml.

Step III

Find the theoretical volume of CO_2 by the decomposition of the mixture (actually $NaHCO_3$).

Let the wt. of $NaHCO_3$ in the mixture $= x$ g

∴ wt. of Na_2CO_3 in the mixture $= (1-x)$ g.

Now the equation tells that

2×84 g of $NaHCO_3$ gives $CO_2 = 22400$ ml

or 22400 ml of CO_2 is obtained from $NaHCO_3 = 2 \times 84$ g

1 ml of CO_2 is obtained from $NaHCO_3 = \dfrac{2 \times 84}{22400}$ g

so 106·9 ml of CO_2 is obtained from $NaHCO_3$

$$= \dfrac{2 \times 84 \times 106.9}{22400} = 0.801 \text{ g}$$

Thus the amount of $NaHCO_3$ in 1 g of the mixture
$= 0.801$ g.

∴ wt. of Na_2CO_3 in the mixture $= 1 - 0.801 = 0.199$ g.

Percentage of $NaHCO_3 = 80.1$

and Percentage of $Na_2CO_3 = 100 - 80.1 = 19.9$.

Example 10. *How much marble of 96·5% purity would be required to prepare 10 litres of CO_2 at N.T.P. when the marble is acted upon by dilute HCl?*

Step I

Write the equation involved.

$$CaCO_3 + 2HCl \longrightarrow CaCl_2 + H_2O + CO_2.$$

Step II

Find the weight of pure $CaCO_3$.

From the equation we see that

22·4 litres of CO_2 are produced by $CaCO_3 = 100$ g

1 litre of CO_2 is produced by $CaCO_3 = \dfrac{100}{22.4}$ g

Problems Based on Equation (Contd.)

10 litres of CO_2 are produced by $CaCO_3$

$$= \frac{100}{22 \cdot 4} \times 10 = \frac{1000}{22 \cdot 4} = 44 \cdot 64 \text{ g}$$

STEP III

Find the weight of $CaCO_3$ (marble of 96.5% purity).

Since the marble of 96·5% purity is needed which means if we have 96·5 g pure $CaCO_3$ the wt. of impure $CaCO_3$ that can be obtained = 100 g.

Thus 96·5 g pure CaO_3 gives impure $CaCO_3$ = 100 g

1 g pure $CaCO_3$ gives impure $CaCO_3 = \frac{100}{96 \cdot 5}$ g

44·64 g pure CaO_3 gives impure $CaCO_3$

$$= \frac{100}{96 \cdot 5} = 44 \cdot 64 = 46 \cdot 25 \text{ g}$$

Thus the weight of 96·5% marble needed for producing 10 litres of CO_2 gas at N.T.P.

$$= 46 \cdot 25 \text{ g}$$

END-OF-CHAPTER PROBLEMS

1. What volume of hydrogen at N.T.P. is obtained by the action of 5·0 g of zinc metal on dil. HCl ? [**Ans.** 1·723 litres]

2. What volume of chlorine gas is obtained by the action of conc. HCl on 8·7 g manganese dioxide at *N*.T.P. ? [**Ans.** 2·24 litres]

3. What weight of $KMnO_4$ and what volume of HCl (sp. gravity 1·2112) would be required to produce 8·00 litres of chlorine at 15°C and 759 mm pressure. [**Ans.** $KMnO_4$ = 21·37 g HCl = 32·58 ml]

4. What volume of phosphine (PH_3) measured at 16°C and 730 mm pressure would be formed by the decomposition of 18 g of phosphonium iodide by caustic soda ? [**Ans.** 2·743 litres]

5. What volume of CO_2 at 14°C and 730 mm pressure can be obtained by the action of HCl on 20 gm of sodium carbonate ? [**Ans.** 2·626 litres]

6. What weight of ammonium nitrate is required to give 5·6 litres of nitrous oxide at S.T.P. ? [**Ans.** 20·00 g]

7. What weight of sodium chloride is required to give 42 litres of chlorine at S.T.P. ? [**Ans.** 219·0 g NaCl]

8. What volume of CO_2 at 15°C and 750 mm pressure is required to convert 8 g of sodium carbonate into sodium bicarbonate?

[**Ans.** 1806·4 ml]

9. What volume of chlorine measured at 12°C and 780 mm pressure can be obtained when 110 g of manganese dioxide acts upon conc. hydrochloric acid? If the acid contains 1·38% HCl and has specific gravity 1·2, what is the volume of HCl needed?

[**Ans.** 28·8 litres; 404·9 ml]

10. What volume of air containing 21% oxygen by volume is required to completely burn 1 kg of sulphur, 4% out of which is incombustible matter? [**Ans.** 3200 litres]

11. What weight of copper must be treated with dilute nitric acid to give 1·05 litres of nitric oxide at 12°C and 600 mm pressure?

[**Ans.** 3·382 g of Cu]

12. Calculate the amount of zinc necessary to liberate 168 ml of hydrogen at S.T.P. when zinc is treated with dil. H_2SO_4.

[**Ans.** 0·487 g]

13. What volume of oxygen at 17°C and 760 mm pressure would be obtained by heating 100 g of potassium chlorate of 85% purity?

[**Ans.** 24·759 litres]

14. What volume of oxygen at S.T.P., is needed to prepared 11·5 g of phosphorus pentoxide? [**Ans.** 4·63 litres]

15. Calculate the volume of ammonia at S.T.P. required to produce NH_4Cl from 146 g of HCl. [**Ans.** 89·6 litres]

16. What volume of oxygen at 17°C and 770 mm pressure can be obtained from 50 g of a solution containing 10% by weight of hydrogen peroxide? [**Ans.** 1·726 litres]

17. What weight of calcium carbonate will be precipitated from excess of lime water by 0·5 litre of CO_2 at 6°C and 765 mm pressure?

[**Ans.** 2·20 g]

18. 200 g of marble chips are dropped into one kilogram of solution of HCl containing one tenth of its weight of pure acid. How much chips will remain undissolved? What weight of $CaCl_2$ and what weight of CO_2 could be obtained from it?

(*Kurukshetra Pre-Univ., 1977*)

[**Ans.** 63 g; 152·07 g; 60·28 g]

19. Hydrogen reacts essentially completely with CuO to form copper and water. How many grams of hydrogen are needed to produce

(a) 0·4 g of copper? (b) 0·4 mole of copper?

(c) 0·4 g of water?

[Cu = 63·5, O = 16, H = 1]
(*Kurukshetra Pre-Univ., 1977*)

[**Ans.** (a) 0·012 g (b) 0·8 g) (c) 0·044 g]

Problems Based on Equation (Contd.)

20. What volume of oxygen at 18°C and 750 mm pressure can be obtained from 10 g $KClO_3$? (At. wts. K = 39, Cl = 35·5, O = 16)
(Himachal Pre-Univ., 1978)
[**Ans.** 4·23 litres]

21. 448·8 ml of ammonia at N.T.P. is oxidised to nitric oxide as under :

$$4NH_3 + 5O_2 \longrightarrow 4NO + 6H_2O$$

Find out the volume of oxygen needed for oxidation and volume of nitric oxide formed at 0°C and 76 cm pressure.
(Punjab Pre-Univ., 1981)
[**Ans.** Volume of O_2 = 561 ml
Volume of NO = 448·8 ml]

22. Calculate the volume of hydrogen liberated at NTP when 500 ml of 0·5 N sulphuric reacts with zinc. (H = 1, O = 16, S = 32).
(Kurukshetra Pre-Univ., 1981)
[**Ans.** 5·6 litres]

23. 200 gms of Marble chips are dropped into one kilogram of a solution of HCl, containing one tenth of its weight of pure acid. How much of the chips will remain undissolved ? What weight of calcium chloride and what weight of carbon dioxide gas could be obtained from it ? (Ca=40 ; C=12 ; O=16 ; Cl=35·5 ; H=1).
Kurukshetra Pre-Univ., 1982)
[**Ans.** Marble=63·02 gm ; $CaCl_2$=152·04 gm ; CO_2=60·27 gm.]

24. What weight of $KMnO_4$ and what volume of HCl (specific gravity=1·212) would be required to produces 8 litres of chlorine at N.T.P. ? (K=39 ; Mn=55 ; O=16 ; Cl=35·5 ; H=1)
[**Ans.** $KMnO_4$=25·35 gm ; HCl=34·39 cc]

Percentrage Purity

25. A large lump of zinc is placed in 100 g of solution of HCl, 150 ml of hydrogen at 15°C and 740 mm pressure are evolved. Calculate the percentage of HCl in the solution. [**Ans.** 0·451%]

26. 84 litres of H_2S at S.T.P. were liberated from 35·00 g of a sample of iron sulphide on treatment with excess of H_2SO_4. Calculate the percentage purity of FeS in the given sample. [**Ans** 94·2%]

27. One gram of a mixture of $CaCO_3$ and $MgCO_3$ gives 240 cc of CO_2 at S.T.P. Calculate the percentage composition of the mixture,
[**Ans** $CaCO_3$ = 62·5% ; $MgCO_3$ = 37·5%]

28. How much marble of 90% purity would be required to prepare 4·5 litres of CO_2 at N.T.P. when the marble is acted upon by dilute hydrochloric acid ? [**Ans.** 22·29 g of marble]

29. What volume of oxygen measured at S.T.P. will be obtained by heating 0·85 g sodium nitrate ? [**Ans.** 0·56 litre]

30. Calculate the weight of potassium chlorate required to produce 336 litres of oxygen at N.T.P. *(Punjab H.S., 1970)*
[**Ans.** 12·25 gm.]

31. What weight of hydrochloric acid will be required to completely dissolve 5 gm. of calcium carbonate ? Calculate the volume of carbon dioxide evolved at N.T.P. *(Punjab H.S.1971)*
[**Ans.** 3·65 gm ; 1120]

32. What weight of H_2SO_4 will be required to completely dissolve 4·2 gm. of magnesium carbonate. Calculate the volume of CO_2 evolved at N.T.P.
(Punjab Pre-Univ., 1973)
[**Ans.** 4·9 gm ; 1120 ml.]

33. What volume of CO_2 at N.T.P. will be obtained by complete decomposition of one gram molecule of sodium bicarbonate ?
(Shivaji Pre-Univ., 1975)
[**Ans.** 1200 ml.]

34. 2 g of a sample of marble gave 441·2 ml of CO_2 at 8°C and 752 mm pressure on treatment with hydrochloric acid. What is the percentage purity of $CaCO_3$ in the marble ? [**Ans.** 95%]

35. What weight of potassium permanganate is required to obtain 22·4 litres of chlorine at N.T.P. from an excess of concentrated hydrochloric acid ? [K = 39, Mn = 55, O = 16]
(Madurai Pre-Univ., 1975)
[**Ans.** 63·2 g]

36. 9·7 g of mixture containing carbonates of calcium and magnesium when reacted with excess of dilute hydrochloric acid liberate 2·352 litres of carbon dioxide at N.T.P. Calculate the percentage composition of the mixture. *(Punjab Pre-Univ., 1976)*
[**Ans.** $CaCO_3$ = 46·7% $MgCO_3$ = 53·3%]

37. 5 g of natural sulphur when burnt in air produce 3 litres of SO_2 at N.T.P. Calculate the percentage of pure sulphur in the substance. *(Andhra Pre-Univ., 1976)*
[**Ans.** 85·7]

38. How many grams of sodium chloride are required to prepare one litre of 0·2 normal solution ? *(Punjab Pre-Univ., 1976)*
[**Ans.** 11·7g]

39. 2 litres of hydrogen bromide measured at 15°C and 772 mm pressure were passed into a solution of 10 g of caustic potash in water. The solution was then evaporated to dryness. What would the weight of the solid residue, and what percentage of caustic potash would it contain ?
[H = 1, O = 16, Br = 80, K = 39, and one litre of hydrogen at N.T.P. weighs 0·09 g]
(Kurukshetra Pre-Univ., 1978)
[**Ans.** 11·95 g ; 30·25%]

40. What weight of slaked lime would be required to decompose completely 4 g of ammonium chloride, and what would be the weight of each product ? [O = 16, H = 1, Ca = 40, N = 14]
(Kurukshetra Pre-Univ., 1978)
[**Ans.** $Ca(OH)_2$ = 2·76 g ; NH_3 = 1·27 g
$CaCl_2$ = 4·15 g ; H_2O = 1·35 g]

41. 1·84 gm. of a mixture of $CaCO_3$ and $MgCO_3$ are heated strongly till no further loss of weight takes place. The residue strongly till no further loss of weight takes place. The residue weighs 0·96 gm. Find out the percentage composition of the mixture. (Ca=40, C=12, O=16, Mg=24).

(Maharshi Dayanand Pre-Univ. 1982)
[**Ans.** $CaCO_3$=54·34% ; $MgCO_3$=45·66%]

19

Problems Based on Equations (Contd.)

TYPE III—(Volume-Volume Relationship)

Here in these problems the reactants and the products are all gases. We known from the Gay-Lussac's Law of gaseous volumes that **When gases react with each other they do so in terms of volumes which bear a simple ratio to each other**. All these problems are based on the principle of molar volume of gases *i.e., one gram molecule of each gas occupies 22.4 litres at S.T.P.* The volumes represented by a chemical equation are assumed to be at S.T.P.

Example 1. *What volume of oxygen gas is required to burn 150 ml of carbon monoxide at S.T.P. ?*

STEP I

Write the chemical equation involved.

$$2CO + O_2 \longrightarrow 2CO_2$$

STEP II

Find the volume of oxygen required.

From the equation we see that

2 volumes of carbon monoxide require oxygen = 1 volume

or 2 ml of carbon monoxide require oxygen = 1 ml

150 ml of carbon monoxide require oxygen

$$\frac{1}{2} \times 150 = 75 \text{ ml.}$$

Thus the volume of oxygen required for the complete combustion of 150 ml of CO is 75 ml.

Example 2. *A mixture of 200 ml of methane and 800 ml of oxygen is exploded. What is the volume after explosion temperature being constant at 100°C and pressure at 760 mm?*

STEP I

Write the chemical equation involved (Methane is CH_4)

$$CH_4 + 2O_2 \longrightarrow CO_2 + 2H_2O$$

STEP II

Find the volume of the products after explosion

From the equation we see that

1 volume of CH_4 required oxygen = 2 volumes

or 1 ml of CH_4 required oxygen = 2 ml

∴ 200 ml of CH_4 will require oxygen = 2×200 = 400 ml

Total volume of oxygen = 800 ml

∴ volume of oxygen left unused = 400 ml

From the equation we see that volume of CO_2 formed from 1 volume of methane is 1.

Thus 1 ml of CH_4 gives CO_2 = 1 ml

∴ 200 ml of CH_4 will give CO_2 = 200 ml

Also we find from the equation

that 1 volume of methane gives water = 2 volumes

or 1 ml of CH_4 gives water = 2 ml

∴ 200 ml of methane will give water = 400 ml

Since the temperature of the reaction is kept at 100°C.

∴ This water will be in form of steam.

Thus we find that resulting volume of the products is

(a) unused oxygen = 400 ml

(b) CO_2 product = 200 ml

(c) water ,, = 400 ml

Total volume = 1000 ml

Problems Based on Equations (Contd.)

Example 3. *What volume of oxygen will be evolved by the decomposition of 120 ml of hydrogen peroxide marked '30' volume ?*

'30' volumes of hydrogen peroxide means that 1 ml of such a H_2O_2 will give 30 ml of oxygen at S.T.P.

Thus 100 ml of '30' volume H_2O_2 will require 100×30 *i.e.* 3000 ml oxygen at S.T.P.

$$(2H_2O_2 \longrightarrow 2H_2O + O_2)$$

Example 4. *Calculate the volume of air containing 21% by volume of oxygen at S.T.P. required in order to convert 204 ml of sulphur dioxide to sulphur trioxide under the same conditions.*

Step I

Write the chemical equation involved.

$$2SO_2 + O_2 \longrightarrow 2SO_3$$

Step II

Find the volume of oxygen.

From the equation, it is seen that

2 volumes of sulphur dioxide react with oxygen = 1 volume

or 2 ml of SO_2 react with oxygen = 1 ml

$\therefore$ 204 ml of SO_2 will require oxygen = $\dfrac{204}{2}$

= 102 ml.

Step III

Find the volume of 21% oxygen.

21% oxygen by volume means that in 100 ml of air the oxygen present is only 21 ml.

Thus 21 ml of pure oxygen give air = 100 ml

1 ml of pure oxygen give air = $\dfrac{100}{21}$

102 ml of pure oxygen give air = $\dfrac{100}{21} \times 102$

= 485·7 ml.

Hence the volume of air = 485·7 ml.

Example 5. *A sample of coal gas contains 45% hydrogen, 30% methane, 15% carbon monoxide and 10% acetylene. 100 volume of this gas were mixed with 200 volumes of oxygen and exploded. Find the composition of resulting mixture on cooling.*

STEP I

Write the equations involved in the problem.

$$2H_2 + O_2 \longrightarrow 2H_2O$$
$$CH_4 \text{(Methane)} + 2O_2 \longrightarrow CO_2 + 2H_2O$$
$$2CO + O_2 \longrightarrow 2CO_2$$
$$2C_2H_2 \text{ (acetylene)} + 4O_2 \longrightarrow 5CO_2 + 2H_2O$$

STEP II

Find the volume of oxygen used for combustion of all the component gases.

2 vols. of hydrogen require 1 vol. of oxygen.

$\therefore$ 1 vol. of H_2 will require oxygen = $\dfrac{1}{2}$ volume

and 45 vols. of H_2 will require oxygen = $\dfrac{45}{2}$

= 22·5 volumes

1 volume of methane require oxygen = 2 volumes

$\therefore$ 30 volumes of methane require oxygen = 30 × 2
= 60 volumes

Again 2 volumes of carbon monoxide require oxygen
= 1 volume

$\therefore$ 15 volumes of carbon monoxide require oxygen

= $\dfrac{15}{2}$ = 7·5 volumes.

Similarly 2 volumes of acetylene require oxygen
= 5 volumes

1 volume of acetylene require oxygen

$$= \frac{5}{2} \text{ volumes,}$$

10 vol. of acetylene require oxygen $= \frac{5}{2} \times 10$

$$= 25 \text{ volumes.}$$

Total volume of oxygen required for the complete combustion of 100 volumes of the mixture

$$= 22\cdot 5 + 60 + 7\cdot 5 + 25$$
$$= 115\cdot 00 \text{ volumes}$$

∴ Volume of oxygen left unused

$$= 200 - 115$$
$$= 85 \text{ volumes.}$$

STEP III

Find the resulting volume of each gas after combustion.

In the resulting mixture, unused oxygen and carbon dioxide only are present, steam produced condenses on cooling and as such its volume is not recorded.

From the equation we find that

Vol. of CO_2 produced from 30 vols. $CH_4 = 30$ volumes

Vol. of CO_2 produced from 10 vols. of $CO = 10$ volumes

Vol. of CO_2 produced from 10 vols. of $C_2H_2 = 20$ volumes

∴ Total vol. of $CO_2 = 30 + 10 + 20$

$$= 60$$

Thus in the resulting mixture:

Total CO_2 present = 60 volumes

Oxygen left unused = 85 volumes

Total volume = 145 volumes.

Example 6. *20 ml of a mixture of carbon monoxide and methane required for complete combustion 28 ml of oxygen measured at S.T.P. Find the composition of the mixture.*

STEP I

Write the chemical equations involved.

$$2CO + O_2 \longrightarrow 2CO_2$$
$$CH_4 + 2O_2 \longrightarrow CO_2 + 2H_2O$$

STEP II

Find the total theoretical volume of oxygen required by the mixture.

Let the vol. of CO in the mixtue $= x$ ml

∴ The vol. of CH_4 in the mixture $= 20-x$

From the equation we see that

2 vols. of CO require oxygen $= 1$ vol.

∴ x vols. of CO require oxygen $= \dfrac{x}{2}$

Similarly

1 vol. of CH_4 require oxygen $= 2$ vols.

$(20-x)$ vols. of CH_4 require oxygen $= 2(20-x)$

Total theoretical vol. of oxygen required

$$= \dfrac{x}{2} + 2(20-x)$$

STEP III

Find the value of x.

Actual volume of oxygen consumed by the mixture

$$= 28 \text{ ml}$$

∴ $\dfrac{x}{2} + 2(20-x) = 28$

or $\quad x + 4(20-x) = 56$

or $\quad\quad -3x = -24$

or $\quad\quad\quad x = 8$

∴ The volume of CO in the mixture $= 8$ ml
and the volume of CH_4 in the mixture $= 20 - 8 = 12$ ml.

Problems Based on Equations (Contd.)

Example 7. *15 ml of a mixture of methane (CH_4) and ethylene (C_2H_4) was exploded with oxygen and the volume of CO_2 produced after explosion and cooling was 25 ml. Find the composition of the mixture.*

Step I

Write the chemical equations involved.

$$CH_4 + 2O_2 \longrightarrow CO_2 + 2H_2O$$
$$C_2H_4 + 3O_2 \longrightarrow 2CO_2 + 2H_2O$$

Step II

Find the theoretical volume of CO_2 produced.

Let the vol. of CH_4 in the mixture $= x$ ml

the vol of C_2H_4 in the mixture $= (15-x)$ ml

From the equation we see that

1 vol of CH_4 produces $CO_2 = 1$ vol

∴ x ml of CH_4 produces $CO_2 = x$ ml

Similarly, 1 vol. of C_2H_4 produces $CO_2 = 2$ vols

or   1 ml of C_2H_4 produces $CO_2 = x$ ml

Thus $(15-x)$ ml of C_2H_4 produce $CO_2 = 1$ $(15-x)$ ml

Total theoretical volume of CO_2 produced

$$= x + 2(15-x) \text{ ml}$$

Step III

Find the value of x.

The actual volume of CO_2 produced from 15 ml of mixture
$$= 25 \text{ ml}$$

∴ $x + 2(15-x) = 25$

or $x + 30 - 2x = 25$

or $-x = -5$

or $x = 5$

∴ The volume of CH_4 in the mixture $= 5$ ml
and the vol. of C_2H_4 in the mixture $= 15 - 5 = 10$ ml.

Example 8. *A sample of ozonised oxygen contained 75% by volume of oxygen. What will be the volume of the resulting gases when 100 ml of such a sample is heated and then cooled to original conditions of temperature and pressure.*

Step I

Write the chemical equations involved.

$$2O_3 \longrightarrow 3O_2$$

Step II

Find the volume of ozone in the given sample of ozonised oxygen.

Since oxygen is 75%, it means 25% must be volume of ozone.

Thus 100 ml of ozonised oxygen contain ozone = 25 ml.

Step III

Find the volume of oxygen obtained by the decomposition of 25 ml of ozone.

From the equation we see that

2 vols. of ozone give oxygen = 3 volumes

$\therefore$ 1 vol. of ozone gives oxygen = $\dfrac{3}{2}$ volumes

and 25 ml of ozone gives oxygen = $\dfrac{3}{2} \times 25$

$$= \dfrac{75}{2} = 37 \cdot 5 \text{ ml.}$$

Step IV

Find the total volume of oxygen.

Already 100 ml of the given sample contains 75 ml of oxygen and 37·5 ml of oxygen have been formed from ozone present in this sample.

$\therefore$ Total volume of oxygen = 75 + 37·5 = 112·5 ml.

END-OF-CHAPTER PROBLEMS

1. 20 ml of dry ammonia were completely decomposed by sparking. Find the volumes of nitrogen and hydrogen evolved. Temperature remaining constant. [**Ans.** 10 ml N_2; 30 ml H_2]

Problems Based on Equations (Contd.)

2. What volume of chlorine gas is required for completely neutralizing 16 ml of ammonia in the reaction

$$8NH_3 + 3Cl_2 \longrightarrow N_2 + 6NH_4Cl$$

What volume of nitrogen will be produced? Temperature remaining constant. [**Ans.** 6 ml of Cl_2; 2 ml of N_2]

3. 50 ml of carbon disulphide vapour were mixed with 175 ml of oxygen and the mixture exploded. What is the composition of the product? [**Ans.** 50 ml CO_2; 100 ml SO_2; 25 ml O_2]

4. A sample of ozonised oxygen contained 80% by volume of oxygen. Calculate the volume when 100 ml of the same sample of ozonised oxygen is heated and cooled to original conditions of temperature and pressure. [**Ans.** 110 ml]

5. A sample of ozonised oxygen contained 90% by volume of oxygen. Calculate the volume of the resulting gas when 120 ml of the the same sample of ozonised oxygen is heated and cooled to original conditions of temperature and pressure. [**Ans.** 126 ml]

6. One litre of oxygen at S.T.P. is made to react with three litres of carbon monoxide at S.T.P. Calculate the volume of the resulting gases. [**Ans.** $CO = 1$ litre, $CO_2 = 2$ litres]

7. What volume of CO_2 will be produced by the complete combustion of 20 ml of acetylene (C_2H_2) at S.T.P.? [**Ans.** 40 ml]

8. A sample of coal gas was found to contain hydrogen, methane and carbon monoxide. 30 ml of this mixture was exploded in a eudiometer tube with 120 ml oxygen. On cooling the resulting gas was found to be 120 ml. Absorption with caustic potash further reduced the volume by 15 ml. Calculate the percentage of the constituent in the original mixture.

[**Ans.** $H_2 = 50\%$, $CH_4 = 40\%$, $CO = 10\%$]

9. 30 ml of a mixture containing 9 ml ethane and 21 ml oxygen was exploded. What was the volume of each component of the resulting mixture? (All measurements being made at 100°C and 760 mm pressure). [**Ans** $CO_2 = 12$ ml, $H_2O = 3$ ml; CH_4 left unreacted $= 3$ ml]

10. What volume of oxygen will be evolved by the decomposition of 70 ml of "20 volume" H_2O_2 (at S.T.P.)?

[**Ans.** 1400 ml of oxygen]

11. 10 ml of a mixture of carbon monoxide and methane required for complete combustion 14 ml of oxygen. Find the volume of each component in the mixture.

[**Ans.** $CO = 4$ ml ? $CH_4 = 6$ ml]

12. 20 ml of a mixture of CH_4 and C_2H_4 were decomposed with excess of oxygen. The volume of CO_2 produced after explosion and cooling was 30 ml. Find the percentage composition of the mixture.

[**Ans.** $CH_4 = 50\%$; $C_2H_4 = 50\%$]

13. 15 ml of a mixture of H_2 and C_2H_2 required 19·5 ml of oxygen for complete combustion. Calculate the percentage composition of the mixture. [**Ans.** H_2 = 60% ; C_2H_4 = 40%]

14. 20 cc of a mixture of carbon monoxide and acetylene were fired with 80 cc oxygen. The residual gases occupied 34 cc and after treatment with caustic potash, the residual oxygen occupied 8 cc. What was the composition of the mixture ?
[**Ans.** CO = 14 cc, acetylene = 6 cc]

15. Calculate the volume of oxygen measured at 127°C and 760 mm pressure that will combine with 160 ml of hydrogen at 0°C and 190 mm pressure to form water. [**Ans.** 29·30 ml]

16. What is the volume of oxygen required for complete combustion of one litre of methane at N.T.P. ? *(Sambalpur Pre-Univ., 1973)*
Ans. 2 litres]

17. Calculate the volume of air (containing 21% oxygen by volume) which would be required to completely burn 5 litres of methane (CH_4) at S.T.P. *(Punjab Pre-Univ., 1974)*
[**Ans.** 47·62 litres]

18. Find out the volume of air at 91°C and 700 mm pressure, that would be required to burn completely one kilogram of coal containing 45% carbon and 55% incombusible matter. (Air contains 20% of oxygen by volume). *(Kurukshetra Pre-Univ., 1977)*
[**Ans.** 60·80 litres)

16

Measurement and Uncertainty

Measurements of the quantities of material involved in chemical reactions is a basic procedure in chemistry. It is very important to have exact measurement of various objects. Every particular measurement is expressed in numbers. Numbers used in chemistry problems represent one of the following:

MEASURED VALUES

These are such numerical values that are obtained by making actual measurements with the help of instruments such as a balance or through calculations of the measurements made by the instruments. For example, there are 412 trees in a garden. This number is the result of actual counting of the trees by someone. 412 is thus a measured value. Another example is the value 14·31 ml of water in a beaker. It is also a measured value because the volume of water has been actually measured with the help of a burette or a pipette.

DEFINED VALUES

These are such numerical values which are known by definition For example, 1 dozen beakers. It means 12 beakers. one dozen by definition is equal to 12 articles or 1 litre of water. It means 1000 ml of water. By definition 1 litre is equal to 1000 ml.

Example 1. *Which of the following are measured and defined values?*

(i) *20 eggs per score.*

(ii) The bench is 1·71 m long.

(iii) They ate 20 eggs at the breakfast.

(iv) A litre is equivalent to 1000 ml.

(v) They brought four cups of tea.

Measured values

(ii) 1·71 m;   *(iii)* 20 eggs;   *(v)* 4 cups of tea.

Defined values

(i) 20 eggs per score (by definition);

(ii) a litre is equal to 1000 ml (by definition);

The value 20 of eggs in *(iii)* is obtained by actual counting while 20 eggs per score in *(i)* is a definition of the score.

Accuracy of Measurement and Significant Numbers

Every measurement is the result of some measuring instrument. Therefore, every observed number is subject to both *observational* error and instrumental faults. Some observers have sharper eyes or steadier hands than others. We cannot say, then, that any measurement of a physical quantity is exact. All we can say at least is that it has been made more precisely, with more careful technique with better instruments—than other measurements. The numerical value of every measurement that we make, is an approximation. No physical measurement such as length, mass, volume, velocity, density, time is ever absolutely correct. Thus 5·230 g is a more precise weighing than 5·23 g. The former measurement means that the weight is closer to 5·229 g and the latter makes only a modest claim towards precision.

The precision of a measurement is revealed by the number of **significant numbers or figures**. *A significant figure is one which represents a reliable value of a measure.* scientific measurements are always reported with numbers whose last digits are somewhat in doubt. The value 5·23 g contains two digits which are certain *i.e.*, 5 and 2; and one more digit 3 which is uncertain. These are called significant figures.

HOW TO DETERMINE THE SIGNIFICANT FIGURES ?

Where numbers are added, subtracted, multiplied or divided, the number of significant figures in the answer depends upon the position of decimal point. The answer can have no more significant figures to the right of the decimal point than are contained in the number with fewest digits to the right of the decimal point. For example, we want to find the sum of three numbers 22·06, 0·00739 and 113·5. The sum is

$$20·07$$
$$0·00739$$
$$113·5$$
$$\overline{135·57739}$$

Out of the three numbers given above, one has only one digit to the right of the decimal *i.e.*, 113·5. The number obtained after adding is actually 135·57739. It is wrong to report it as such. The correct answer is 135 6· The number 135 57739 has to be rounded off, so as to contain only one digit to the right of the decimal. The significant figures in the sum are then 4. It should be noted that *zero used to fix the decimal point are not counted*.

Example 2. *How many significant figures are used to express the following value ?*

(*i*) 2·038 ml;   (*ii*) 0·0371 g;   (*iii*) 12·0037 g

(*i*) has **four** significant numbers,

(*ii*) has **three** significant numbers. (zeros used only to fix the decimal point are not counted)

(*iii*) has **six** significant figures.

WHICH VALUE HAS GREATER PRECISION ?

The value which has greater number of significant figures is more precise.

Example 3. *A beaker weighs 20·23 g in an ordinary balance, while it weighs 20·2292 g on an analytical balance. Which measurement has greater precision ?*

The second measurement *i.e.* 20·2292 g is more accurate or precise since it has larger number (6) of the significant number in comparison to the first (which has only 4).

Example 4. *Express the following numbers with an accuracy of three significant figures : 296·8, 520·5, 421·5 and 422·5.*

The number 296·8 should be rounded off so as to get three significant numbers. It is 297. The second is 520 but since the number is midway between 520 and 521, an even number is preferred as a convention. Hence the answer is 520. The value for third and fourth numbers are both rounded off to 422.

(i) $\quad 2\cdot 5 \left(\dfrac{27\cdot 2}{3\cdot 27} \right)$

(ii) $\quad 10\cdot 2 \left(\dfrac{52\cdot 7}{26\cdot 2} \right) \left(\dfrac{1\cdot 052}{99\cdot 8} \right)$

(i) $\quad 2\cdot 5 \left(\dfrac{27\cdot 2}{3\cdot 27} \right) = 2\cdot 5 \times \dfrac{27\cdot 2}{3\cdot 27} = 21$

Since among all the numbers involved, the one with fewest significant number is 2 (the no. 2·5 has 2 significant numbers), the answer 21 is correct since it has 2 significant figures.

(ii) $\quad 10\cdot 2 \left(\dfrac{52\cdot 7}{26\cdot 2} \right) \left(\dfrac{1\cdot 052}{99\cdot 8} \right) = 0\cdot 0216$

The answer should have only three significant figures because among the number involved in the above calculations one has three significant numbers (10·2). Hence the 0·0216 is correct since it has three significant numbers.

Example 5. *How many significant numbers are involved in the number of milligrams when we say "A gram has one thousand milligram".*

Now, 1 gram = 1000 milligram.

It is a defined value. The number of significant numbers in a defined value is always infinity.

Expressed mathematically as

$$1 \text{ gram} = 1{,}000\cdot 00\ldots\ldots \text{mg}.$$

EXPRESSING UNCERTAINTY

The last significant number is always uncertain. For example, we express the length of a wooden piece as 124 mm. The last number is always in doubt. Correct value may be between 123·5 and 124·5 mm. The digits 1 and 2 in 124 are certain, while the degree of uncertainty is expressed by the last digit *i.e.* 4. The number, which is uncertain, is expressed as a

Measurement and Uncertainty

± quantity. Thus the number 124 may be expressed as 123 $\pm \triangle$ A. The quantity $\pm \triangle$ A is the uncertainty in measurment. If we have two quantities each may be expressed in terms of uncertainties $\pm \triangle$ A and $\pm \triangle$ B. Suppose we have weighed three different substances A, B and C, the values of their weights are best expressed as (say). $A = 11.25 \pm 0.20$; $B = 25.75 \pm 0.25$ g and $C = 17.85 \pm 0.53$ g. The sum of all the three is

$$A = 11.25 \pm 0.20 \ g$$
$$B = 25.75 \pm 0.25 \ g$$
$$C = 17.85 \pm 0.53 \ g$$
$$\overline{A+B+C = 54.85 \pm 0.98 \ g}$$

Example 6. *The volumes of three different substances measured by three different observers are*

$$A = 47.15 \pm 0.22 \ ml$$
$$B = 13.25 \pm 0.17 \ ml$$
$$C = 21.35 \pm 0.20 \ ml$$

Find the value of $2A + 2B + 3C$.

$2A = 2 \ (47.15 \pm 0.22) \ ml = 94.30 \pm 0.44 \ ml$
$2B = 2 \ (13.25 \pm 0.17) \ ml = 26.50 \pm 0.34 \ ml$
$3C = 3 \ (21.35 \pm 0.20) \ ml = 64.05 \pm 0.60 \ ml$

Thus, $2A + 2B + 3C$ is

$$94.30 \pm 0.44 \ ml$$
$$26.50 \pm 0.34 \ ml$$
$$64.05 \pm 0.60 \ ml$$
$$\overline{184.85 \pm 1.38 \ ml}$$

Example 7. *An object weighs 40.2 ± 0.1 g and is moving with a velocity of 10.2 ± 0.2 cm per sec. Find the momentum.*

Momentum = Mass × velocity.

$$= (40.2 \pm 0.1) \ (10.2 \pm 0.2)$$

STEP I

Find the percentage uncertainties of mass and velocity.

Percentage uncertainty in mass $= \dfrac{0.1 \times 100}{40.2} = 0.024 = 0.02$

Percentage uncertainty in velocity $= \dfrac{0 \cdot 2 \times 100}{10 \cdot 2} = 2\%$

STEP II

Find Momentum.

$$\therefore \text{Momentum} = (40 \cdot 2 \pm 0 \cdot 02\%)(10 \cdot 2 \pm 2\%)$$
$$= (40 \cdot 2 \times 10 \cdot 2) \pm (0 \cdot 02 \times 2)$$
$$= 410 \cdot 04 \pm 0 \cdot 04\%$$

STEP III

Change Percentage to Uncertainly.

$$\text{Momentum} = 410 \cdot 04 \pm \dfrac{0 \cdot 04}{100}$$

$$= 410 \cdot 04 \pm 0 \cdot 0004.$$

END-OF-CHAPTER PROBLEMS

1. Which of the following are measured and defined values ?

    (*i*) There are 219 plants in a plot

    (*ii*) They brought 6 cups of tea

    (*iii*) A score is equal to 20 articles

    (*iv*) There is 40·2 ml of water in a flask

    (*v*) A gram has 1000 milligrams.

    [**Ans.** *Measured* (*i*); (*ii*), (*iv*), *Defined* (*iii*), (*v*)]

2. How many significant numbers are used to express the following :

    (*i*) 2·078

    (*ii*) 0·0029

    (*iii*) 1·00678   [**Ans.** (*i*) 4, (*ii*) 2, (*iii*) 6]

3. Find the number that will represent the sum of 39·32 and 0·5432.   [**Ans.** 39·86]

4. Find the number that will represent the subtraction of 12·076 from 112·7.   [**Ans.** 100·6]

5. How many significant numbers should the answer of this calculation contain ?

$$2 \cdot 413 + \dfrac{1}{1000}$$ [**Ans.** Four]

6. How many significant figures are there in

(i) 100±1 cc (ii) 100+10 cc (iii) 100×10² (iv) 1·0×10² cm
(v) 1×10² cc.   [**Ans.** (i)=3 (iii)=3 (v) 1
                            (ii)=2 (iv)=2]

7. Write a short note on methods of communicating scientific regularities.   (*Punjab Pre-Univ., 1973*)

8. Calculate the maximum uncertainty in each of the following :

(i) 51·2±0·2°C   (ii) 500·00±0·2 ml
   −23·4±0·2°C   +10:0±0·1 ml
                                (*Punjab Pre-Univ., 1973*)
        [**Ans.** (i) 27·8+0·4°C; (ii) 510·0±0·3 ml]

9. Write a short note on the importance of *measurement* in every day life.   (*Punjab Pre-Univ., 1974*)

17

Glossary

Acid (*Arrhenius definition*). A substance that yields hydrogen ions (H+) when dissolved in water.

Acid (*Bronsted-Lowry definition*). A substance that can give or donate a proton (H+) to some other substance.

Anions. Ions with a negative charge.

Atom. The smallest particle of an element that can undergo chemical changes in a reaction.

Atomic Mass Scale. Relative scale of atomic masses, based on an arbitrarily assigned value of exactly 12 atomic mass units (amu) for the mass of carbon—12.

Atomic Number. Number of protons found in the nucleus of an atom of an element.

Avogadro's Number (N) 6.02×10^{23}. The number of particles such as atoms, formula units, molecules, or ions that constitute one mole of the said particle.

Base (*Arrhenius*). A substance that yields hydroxide ions (OH-) when dissolved in water.

Base (*Bronsted-Lowry*). A substance capable of receiving or accepting a proton (H+) from some other substance.

Boiling Point. The temperature at which the vapor pressure of the liquid is equal to the external pressure acting upon the surface of the liquid. The normal boiling point of a liquid is the temperature at which the vapor pressure of the liquid is 760 mm.

Bond Angle. The angle formed between three atoms in a molecule.

Bond Length. The distance between the nuclei of covalently bonded atoms.

Boyle's Law. At constant temperature, the volume of a fixed mass of a given gas is inversely proportional to the pressure it exerts.

Calorie. The amount of heat required to raise the temperature of 1·00 g of water from 14·5 to 15·5°C.

Catalyst. A substance that speeds up a chemical reaction but is recovered without appreciable change at the end of the reaction.

Cations. Ions with a positive charge.

Charles' Law. At constant pressure, the volume of a fixed mass of a given gas is directly proportional to the Kelvin (absolute) temperature.

Chemical Changes. Changes in substances that can be observed only when a change in the composition of the substance is occuring. New substances are formed.

Chemical Properties. Properties of substances that can be observed only when a substance undergoes a change in composition.

Compound. A pure substance that can be broken down by various chemical means into two or more different substances.

Conservation of Energy, Law of. Energy can be neither created nor destroyed, but may be transformed from one form to another.

Conservation of Mass, Law of. Mass can be neither created nor destroyed.

Coordinate Covalent Bond. Formed when both of the electrons of the electron-pair bond are supplied by one atom.

Covalent Bond. Formed by the sharing of electrons between atoms.

Dalton's Law of Partial Pressure. Each gas in a mixture of gases exerts a partial pressure equal to the pressure it would exert if it were the only gas present in the same volume; the total pressure of the mixture is then the sum of the partial pressures of all the gases present.

Definite Proportions or Constant Composition, Law of. A given pure compound always contains the same elements in exactly the same proportions by mass.

Deliquescent Substance. A substance that absorbs enough moisture from the air to form a solution, such as calcium chloride ($CaCl_2$).

Density. Mass of a substance occupying a unit volume :

$$\text{Density} = \frac{\text{Mass}}{\text{Volume}}$$

Dissociation. A process referring to the separation of substances into ions by the action of the solvent.

Efflorescent substance. A hydrate that loses its water of hydration when simply exposed to the atmosphere, such as washing soda ($Na_2CO_3.10H_2O$).

Electrolytes. Substances whose aqueous solutions conduct an electric current, as observed by the glowing of a standard lightbulb, because they release ions in the solution.

Electron. Particle having a relative unit negative charge (actual chagre $= -1.602 \times 10^{-19}$ colulomb) with a mass of 2.109×10^{-28} g or 5.486×10^{-4} amu (relatively negligible).

Electrovalent or ionic bond. Formed by the transfer of one or more electrons from one atom to another.

Element. A pure substane that cannot be decomposed into simpler substance by ordinary chemical means. All of its atoms have the same atomic number.

Empirical formula. The formula of a compound that contains the smallest integral ratio of atoms present in a molecule or formula unit of a compound.

Endothermic reaction. A reaction in which heat is absorbed.

Enegry. The capacity for performing work.

Equation, chemical. A shorthand way of expressing a chemical change (reaction) in terms of symbols and formulas.

Evaporation. The actual escape of moleculss (the most energetic) from the surface of the liquid (below the boiling boiling point) to form a vapor in the surrounding space above the liquid.

Exothermic reaction. A reaction in which heat is evolved.

Formula unit. Generally the smallest combination of charged particles (ions) in which the opposite charged present balance each other so that the overall compund has a net charge of zero, such as NaCl.

Gay-Lussac's law. At constant volume, the pressure of a fixed mass of a given gast is directly proportionl to the Kelvin (absolute) temperature.

Gay-Lussac's law of combining volumes. At the same temperature and pressure, whenever gases react or gases are formed they do so in the ratio of small numbers by volume.

Heat of reaction. The number of calories of heat energy evolved or absorbed in a given chemical reaction per given amount of reactants and/or products.

Heterogeneous matter. Matter not uniform in composition and properties, and consisting of two or more physically distinct portions or phases unevenly distributed.

Homogeneous matter. Matter uniform in composition and properties throughout.

Homogeneous mixture. Matter homogeneous throughout, but composed of two or more pure substances whose proportions my be varied without limit.

Hydrates. Crystalline substance that contain chemically bound water in definite proportions. An example is Epsom salts, magnesium sulfate heptahydrate ($MgSO_4 \cdot 7H_2O$).

Hydrogen bond. A type of bond resulting when a hydrogen atom bonded to a highly electronegative atom (F, O, and N) becomes bonded additionally to another electronegative atom. An example is water :

$$\text{H}-\overset{..}{\text{O}}:--\text{H}-\overset{..}{\text{O}}--\text{H}-\overset{..}{\text{O}}$$
$$\phantom{\text{H}-\text{O}:--}|||$$
$$\phantom{\text{H}-\text{O}:--}\text{H}\text{H}\text{H}$$

Hygroscopic substance. A substance that readily absorbs moisture from air, such as sugar.

Indicators. Compounds whose color is affected by acids and bases.

Ionization. A process referring to the formation of ions from atoms or molecules by the transfer of electrons.

Ions. Charged species (atoms or groups of atoms) with positive or negative oxidation numbers.

Mass. The quantity of matter in a particular body.

Mass action, law of The rate of a chemical reaction is proportional to the "active masses" of the reactants. The "active masses" have been found related to the molar concentration of the reactants in moles per liters for solutions or pressure units for gases.

Mass number. Sum of the number of protons and neutrons in the nucleus of an atom of an element.

Matter. Anything that has mass and occupies space.

Melting point (*freezing point*). The temperature at which the liquid and solid forms are in dynamic equilibrium with each other. At dynamic equilbrium, the rate of melting is equal to the rate of freezing.

Mixture. Matter composed of two or more substances, each of which retains its identity and specific properties.

Molality (m). The concentration of solute in a solution expressed as the number of moles of solute per *kilogram of solvent*:

$$m = \text{Molality} = \frac{\text{Moles of solute}}{\text{Kilograms of solvent}}$$

Molarity (M). The concentration of solute in a solution expressed as the number of moles of solute per *liter of solution*:

$$M = \text{Molarity} = \frac{\text{Moles of solute}}{\text{Litre of } \textit{solution}}$$

Glossary

Molar volume of a gas. The volume occupied by one mole of any gas, 22.41 of gas molecules at 0°C and 760 mm.

Mole. The amount of a substance containing the same number of particles, such as atoms, formula units, molecules, or ions, as there are atoms in exactly 12 g of carbon—12. One mole of particles consists of 6.02×10^{23} particles, such as atoms, formula units, molecules, or ions, and this number of particles has a mass equal to the atomic, molecular, or formula mass of the particles expressed in grams.

Molecular formula. A formula composed of an appropriate number of symbols of elements representing one molecule of the given compound. Also defined as the true formula and containing the actual number of atoms of each element in one molecule of the compound.

Molecule. Generally the smallest particle of a pure substance (element or compound) that can exist and still retain the physical and chemical properties of the substance, such as O_2 and H_2O.

Non-electrolytes. Substances whose aqueous solutions do not conduct an electric current. Examples of nonelectrolytes are sugar (sucrose, $C_{12}H_{22}O_{11}$), ethyl alcohol (C_2H_6O), and water (H_2O).

Normality (N). The concentration of a solute in a solution expressed as the number of equivalents of solute per *litre of solution* :

$$N = \text{Normality} = \frac{\text{Equivalents of solute}}{\text{Litre of solution}}$$

Orbital. A region of space within an atom in which there can be no more than two electrons.

Oxidation. A chemical change in which a substance loses electrons or one or more elements in it increase in oxidation number.

Oxidising agent. The substance reduced.

Percent by mass. The concentration of a solute in a solution expressed as parts by mass of solute per 100 parts by mass of solution :

$$\text{Percent by mass} = \frac{\text{Mass of solute}}{\text{Mass of solution}} \times 100$$

Physical changes. Changes in substances that can be observed without a change taking place in the composition of the substance.

Physical properties. Properties of substances that can be observed without changing the composition of the substance.

Polyatomic ions. Ions consisting of two or more atoms with a net negative or positive charge on the ion.

Pressure. Force per unit area.

Proton. Particle having a relative unit positive charge (actual charge $+1.602 \times 10^{-19}$ Coulomb) and with a mass of 1.6725×10^{-24} g or 1.0073 amu (approximately 1 amu).

Reaction rate. The rate or speed at which the products are produced or the reactants consumed in a given reaction.

Reaction, combination. Two or more substances (either elements or compounds) react to produce one substance :

$A+Z \longrightarrow AZ$, where A and Z are elements or compounds

Reaction, decomposition. One substance undergoes a reaction to form two or more substances :

$AZ \longrightarrow A+Z$, where A and Z are elements or compounds

Reaction, replacement. One element reacts by replacing another element in a compound.

1. A metal replacing a metal ion in its salt :

$$A+BZ \longrightarrow AZ+B$$

2. A nonmetal replacing nonmetal ion in its salt :

$$X+BZ \longrightarrow BX+Z$$

Reducing agent. The substance oxidized.

Reduction. A chemical change in which a substance gains electrons, or one or more elements in it decreases in oxidation number.

Salt. A compound formed when one or more of the hydrogen ions of an acid is replaced by a cation (metal or positive polyatomic ion), or when one or more of the hydroxide ions of a base is replaced by an anion (nonmetal or negative polyatomic ion).

Saturated solution. A solution that is in dynamic equilibrium with undissolved solution ($\rightleftharpoons$), that is the rate of dissolution of undissolved solute is equal to the rate of crystallization of dissolved solute, as shown:

$$\text{Undissolved solute} \underset{\text{rate of crystallization}}{\overset{\text{rate of dissolution}}{\rightleftharpoons}} \text{Dissolved solute}$$

Solute. The component of a solution that is in lesser quantity.

Solution. Homogeneous matter composed of two or more pure substances whose composition can be varied within certain limits.

Solvent. The component of a solution that is in greater quantity.

Specific gravity. Density of a substance divided by the density of some substance taken as a standard, usually water at 4°C:

$$\text{Specific gravity} = \frac{\text{Density of substance}}{\text{Density of water at 4°C}}$$

That is, the ratio of the density of the substance to that of the standard.

Specific heat. The number of calories required to raise the temperature of 1·00 g of a substance by 1·00°C.

Structural formula. Formula showing the arrangement of atoms within a molecule, using a dash (—) for each pair of electrons shared between atoms.

Sublimation. The direct conversion of a solid to the vapour without passing through the liquid state.

Substance, pure. Homogeneous matter characterized by definite and constant composition, and definite and constant properties under a given set of conditions.

Supersaturated solution. A solution in which the concentration of solute is greater than that possible in a saturated (equilibrium) solution under the same conditions. This solution is unstable and will revert to a saturated solution if a "seed" crystal of solute is added; the excess solute crystallizes out of solution.

Surface tension. The property of a liquid that tends to draw the surface molecules into the body of the liquid, and hence to reduce the surface to a minimum.

Titration (*with reference to neutralization*). A process for determining the concentration of an acid or base in a solution through the addition of a base or an acid of known concentration, respectively, until the neutralization point or end-point is reached as shown by an indicator or by an instrument such as the pH meter.

Unsaturated solution. A solution in which the concentration of solute is less than that of the saturated (equilibrium) solution under the same conditions.

Valence. A whole number used to describe the combining capacity of an element in a compound.

Vapour pressure. The pressure exerted by the molecules in the vapour (at constant temperature) in dynamic equilibrium with the liquid in a closed system. Dynamic equilibrium is established when the rate of molecules leaving the surface of the liquid (evaporation) is equal to the rate of the molecules re-entering the liquid (condensation).

Weight. The gravitational force of attraction between the body's mass and the mass of the planet or satellite on which it is weighed.

APPENDICES

APPENDIX 1

Average Relative Chemical Atomic Weights Based on C = 12 amu

(Values in brackets denote the mass numbers of most stable known isotopes.)

	Symbol	Atomic Number	Atomic Weight
Actinium	Ac	89	[227]
Aluminium	Al	13	26·9816
Americium	Am	95	[243]
Antimony	Sb	51	121·75
Argon	Ar	18	39·948
Arsenic	As	33	74·9216
Astatine	At	85	[210]
Barium	Ba	56	137·34
Berkelnium	Bk	97	[247]
Beryllium	Be	4	9·0122
Bismuth	Bi	83	208·980
Boron	B	5	10·811
Bromine	Br	35	97·909
Cadmium	Cd	48	112·40
Calcium	Ca	20	40·02
Californium	Cf	28	[249]
Carbon	C	6	12·01115
Cerium	Ce	58	140·12
Cesium	Cs	55	132·905
Chlorine	Cl	17	35·453
Chromium	Cr	24	51·996
Cobalt	Co	27	58·9352
Copper	Cu	29	63·54
Curium	Cm	96	[245]
Dysprosium	Dy	66	162·50
Einsteinium	Es	99	[254]
Erbium	Er	68	167·26
Europium	Eu	63	151·96
Fermium	Fm	100	[252]
Fluorine	F	9	18·9984

Appendix 1

	Symbol	Atomic Number	Atomic Weight
Francium	Fr	87	[223]
Gadolinium	Gd	64	157·25
Gallium	Ga	31	69·72
Germanium	Ge	32	72·59
Gold	Au	79	196·967
Hafnium	Hf	32	179·48
Helium	He	2	4·0026
Holmium	Ho	67	164·930
Hydrogen	H	1	1·00797
Indium	In	49	114·82
Iodine	I	53	126·9044
Iridium	Ir	77	192·2
Iron	Fe	26	55·807
Krypton	Kr	36	83·80
Lanthanum	La	57	138·91
Lawrencium	Lw	103	[257]
Lead	Pb	82	207·19
Lithium	Li	3	6·939
Lutetium	Lu	71	174·97
Magnesium	Mg	12	24·312
Manganese	Mn	25	54·9380
Mendelevium	Md	101	[256]
Mercury	Hg	80	200·59
Molybdenum	Mo	42	95·94
Neodymium	Nd	60	144·24
Neon	Ne	10	20·183
Neptunium	Np	93	[237]
Nickel	Ni	28	58·71
Neobium	Nb	41	92·906
Nitrogen	N	7	14·0067
Nobelium	No	102	[254]
Osmium	Os	76	190·2
Oxygen	O	8	15·9994
Palladium	Pb	46	106·4
Phosphorus	P	15	30·9738
Platinum	Pt	78	195·09
Plutonium	Pu	94	[244]
Polonium	Po	84	[210]
Potassium	K	19	39·102
Praseodymium	Pr	59	140·907
Promethium	Pm	61	[145]
Protactinium	Pa	91	[231]
Radium	Ra	88	[226]
Radon	Rn	86	186·2
Rhenium	Re	75	102·905
Rhodium	Rh	45	85·47
Rubidium	Rb	37	

	Atomic	Atomic Number	Atomic Weight
Ruthenium	Ru	44	101·07
Samarium	Sm	62	150·35
Scandium	Sc	21	44·956
Selonium	Se	34	78·96
Silicon	Si	15	28·086
Silver	Ag	47	107·870
Sodium	Na	11	22·9898
Strontium	Sr	38	87·62
Sulphur	S	16	32·064
Tantalum	Ta	73	180·948
Technetium	Tc	43	[99]
Tellurium	Te	52	127·60
Terbium	Tb	65	158·924
Thalium	Tl	81	204·37
Thorium	Th	20	232·038
Thallium	Tm	69	164·934
Tin	Sn	50	118·69
Titanium	Ti	82	47·90
Tungsten	W	74	183·85
Uranium	U	92	238·02
Vanadium	V	23	50·942
Xenon	Xe	54	131·30
Ytterbium	Yb	70	173·04
Yttrium	Y	39	88·905
Zinc	Zn	20	65·37
Zirconium	Zr	40	91·22

APPENDIX 2

Molecular and Equivalent Weights of Some Substances

Substance	Molecular Weight	Equivalent Weight
Hydrochloric acid, HCl	36·5	36·5
Sulphuric acid, H_2SO_4	98·0	49·0
Nitric acid, HNO_3	63·0	63·0
Oxalic acid, $H_2C_2O_4$	90·0	45·0
Oxalic acid dihydrate, $H_2C_2O_4.2H_2O$	126·0	63·0
Sodium carbonate, Na_2CO_3	106·0	53·0
Sodium bicarbonate, $NaHCO_3$	84·0	84·0
Sodium hydroxide, NaOH	40·0	40·0
Potassium carbonate, K_2CO_3	138·0	69·0
Potassium bicarbonate, $KHCO_3$	100·0	100·0
Potassium hydroxide, KOH	56·0	56·0

APPENDIX 3

Vapour Pressure of water at Different Temperatures

Temp. °C	V.P. (mm of Hg)	Temp. °C	V.P. (mm of Hg)
−5	3.0	26	25.2
0	4.6	27	26.7
1	4.9	28	28.4
2	5.3	29	30.0
3	5.7	30	31.8
4	6.1	35	42.2
5	6.5	40	55.3
6	7.0	45	71.9
7	7.5	50	92.5
8	8.1	55	118.0
9	8.6	60	149.4
10	9.2	65	147.5
11	9.8	70	233.7
12	10.5	75	289.1
13	11.2	80	355.1
14	12.0	85	433.6
15	12.8	90	525.8
16	13.6	91	546.1
17	14.5	92	567.0
18	15.5	93	588.6
19	16.5	94	613.9
20	17.5	95	633.9
21	18.7	96	657.6
22	19.8	97	682.1
23	21.1	98	707.3
24	22.4	99	733.2
25	23.8	100	760.9

APPENDIX 4

Conversion Factors

LENGTH :

    2·54 cm   = 1 inch
    1 metre   = 39·4 inches

MASS :

    453·5 grams   = 1 pound
    1 Kilogram   = 2·20 lb
    28·3 g   = 1 ounce

VOLUME :

    1 mililitre   = 1 cm^3
    1 litre   = 1000 ml
    1 litre   = 1.06 quarts
    28·6 ml   = 1 fluid ounce

TEMPERATURE :

$$\text{Absolute Zero (K)} = -273 \cdot 16°C$$
$$°K = °C + 273 \cdot 16$$
$$°F = \frac{9}{5} °C + 32$$
$$°C = \frac{5}{9} (°F - 32)$$

APPENDIX 5

University Questions, 1982
(Practice Problems)

Chapter 2. Law of Chemical Combination

1. Hydrogen sulphide (H_2S) contains 94·11% ; water (H_2O) contains 11·11% H_2 ; and sulphur dioxide (SO_2) contains 50% O_2. Show that the results are in agreement with law of reciprocal proportion. (*Kurukshetra Pre-Univ., 1982*)

Chapter 4. Mole Concept

2. Calculate the following :

(a) The number of moles of carbon dioxide which contain 8·00 gm of oxygen.

(b) The number of molecules of methane in 0·80 gm of methane.

(c) The weight of 0·25 gm. atoms of calcium.

(d) The weight of molecule of nitrogen in gms.

(e) The volume occupied by 1·40 gm of nitrogen at N.T.P. (*Guru Nanak Dev Pre-Univ., 1972*)

[**Answer.** (a) 0·25 ; (b) 3×10^{21} ; (c) 10 gm ;
(d) $4·65 \times 10^{-23}$ gm ; (e) 1120 cc.]

3. Calculate the number of moles in (a) $6·023 \times 10^{22}$ molecules of nitrogen ; (b) 220 ml of oxygen at N.T.P.
(*Panjab Pre-Univ., 1982*)

[**Answer.** (a) 0·1 ; (b) 0·00098]

4. 18 gm of glucose ($C_6H_{12}O_6$) is dissolved in 900 gm of water. Find out the mole fractions of glucose and water in the solution. (*Panjab Pre-Uuiv., 1982*)

[**Answer.** 0·1 ; 50]

252

Chapter 6. Gas Laws

5. A gaseous mixture containing 0·025 moles of nitrogen and 16·00 gms of oxygen is contained in a flask of five litres capacity at 27°C. Calculate : (a) the partial pressure of each gas ; (b) the total pressure of gaseous mixture.
(*Guru Nanak Dev Pre-Univ., 1982*)

[**Answer.** (a) $N_2 = 1·12$ atmospheres, $O_2 = 2·3$ atmospheres; (b) 3·42 atmospheres].

Chapter 7. Diffusion of gases

7. Two gases A and B having same volume diffuse through a porous pot in 20 seconds and 10 seconds respectively. The molecular weight of (A) is 49. Calculate the molecular weight of (B). (*Kurukshetra Pre Univ., 1982*)

[**Answer.** 196]

8. Two gases, CO_2 and unknown gas, having same volume diffuse through a porous partition in 14 and 10 seconds. Calculate the molecular weight of unknown gas. (At. Wts. $C = 12$, $O = 16$). (*Maharshi Dayanand Pre-Univ., 1982*)

[**Answer.** 86·24]

9. 112 ml of hydrogen diffuse in the same time as 28 ml of an unknown gas. Find the molecular weight of the gas, if that of hydrogen be 2. (*Punjabi Pre-Univ., 1982*)

[**Answer. 64**]

10. Relative densities of carbon dioxide and oxygen are 22 and 16 respectively. If 25 cc of carbon dioxide diffuse in 75 seconds, what volume of oxygen will diffuse in 96 seconds under similar conditions ? (*Punjab Pre-Univ., 1982*)

[**Answer.** 37·44 cc]

Chapter 8. Molecular Weights

11. What will be the weight of 150 cc of oxygen collected over water at 15°C and 740 mm pressure ? (Aq. Tension at 15°C = 13 mm). (*Punjabi Pre-Univ., 1982*)

[**Answer.** 0·194 gm.]

12. 0·150 gm. of a volatile substance, when treated in Victor Meyer's apparatus, displaced 40·50 cc of air collected over water at 15°C and 746 mm pressure. Calculate the molecular weight of the substance. (Aq. Tension at 15°C=13·7 mm).

(*Maharshi Dayanand Pre-Univ., 1982*)

[**Answer.** 90·83]

Chapter 9. Equivalent Weights

13. How many gram equivalents of oxygen are present in 14 litres of oxygen at N.T.P. (*Punjab Pre-Univ., 1982*)

[**Answer.** 1·25]

14. 1·11 gm of the chloride of a metal dissolved in water were treated with excess of silver nitrate solution. The weight of silver chloride was found to be 2·87 gm. Calculate the equivalent weight of the metal. (Eq. Wt. of Ag.=108; Cl=35·5). (*Punjabi Pre-Univ., 1982*)

[**Answer.** 55·5]

15. 3·54 gm of a metal were dissolved in excess of concentrated HNO_3 and the nitrate thus formed was ignited carefully. The weight of the oxide formed was 4·44 gm. Calculate the equivalent weight of the metal. (*Kurukshetra Pre-Univ., 1982*)

[**Answer.** 31·46]

16. 0·637 gm of acid required 21·6 ml of seminormal NaOH for complete neutralisation. Calculate the equivalent weight of acid. (*Punjab Pre-Univ., 1982*)

[**Answer.** 58·98]

17. 25 ml of a decinormal solution of an acid neutralises exactly 20·00 ml of the solution of a base, containing 2·40 gm of the base dissolved per 500 ml of the solution. Calculate the equivalent weight of the base.

(*Guru Nank Dev Pre-Univ., 1982*)

[**Answer.** 38·40]

Chapter 13. Problems Based on Equations

18. 3·68 gm of a mixture of calcium carbonate and magnesium carbonate on heating to a constant weight produced 1·92 gm of residue. Find the percentage composition of the given mixture. (Atomic masses: H=1, C=12, O=16, Ca=40, Mg=24). (*Punjab Pre-Univ., 1982*)

[**Answer.** $CaCO_3$=36·14%; $MgCO_2$=63·86%]

19. Current market prices of Al, Zn, and Fe scraps per kg. are Rs. 30/-, Rs. 24/-, and Rs. 4/- respectively. If hydrogen is to be prepared by the reaction of one of the metals with dilute H_2SO_4, which would be the cheapest and which would be expensive metal ? (At. wts. : Al=27, Zn=65, Fe=56).

(Punjabi Pre-Univ., 1982)

[**Answer**. Cheapest would be Fe ; Expensive would be Zn]

Chapter 14. Problems Based on Equations

20. An aqueous solution prepared by dissolving 90 gms of a dibasic acid (M. Wt. 90) per litre of a solution has a density of 0·99. Calculate normality, molarity and molality of the solution. *(Guru Nanak Dev Pre-Univ., 1982)*

[**Answer**. Normality=2 ; Molarity=1 ; Molality=0·99]

21. 1·84 gm of a mixture of $CaCO_3$ and $MgCO_3$ are heated strongly till no further loss of weight takes place. The residue weighs 0·96 gm. Find out the percentage composition of the mixture. (Ca=40, C=12, O=16, Mg=24).

(Maharshi Dayanand Pre-univ., 19 82)

[**Answer**. $CaCO_3$=54·34% ; $MgCO_3$=45·66%]

22. 200 gms of Marble chips are dropped into one kilogram of a solution of HCl, containing one tenth of its weight of pure acid. How much of the chips will remain undissolved ? What weight of calcium chloride and what weight of carbon dioxide gas could be obtained from it ? (Ca=40; C=12; O=16, Cl=35·5; H=1). *(Kurukshetra Pre-Univ., 1982)*

[**Answer**. Marble=63·02 gm ; $CaCl_2$=152·04 gm ; CO_2=60·27 gm.]

23. What weight of $KMnO_4$ and what volume of HCl (specific gravity=1·212) would be required to produce 8 litres of chlorine at N.T.P. ? (K=39 ; Mn=55 ; O=16 ; Cl=35·5 ; H=1). *(Kurukshetra Pre-Univ., 1982)*

[**Answer**. $KMnO_4$=25·35 gm ; HCl=34·39 cc]

APPENDIX 6

University Questions, 1983
(Practice Problems)

Chapter 2. Laws of Chemical Combination

1. A metal forms two oxides. The higher oxide contains 80% metal. 0·72 gm of a lower oxide gave 0·8 gm of higher oxide on oxidation. Calculate the weights of oxygen that combine with fixed weight of the metal in two oxides. Also show that the data illustrate the law of multiple proportions.
(Punjab Pre-Univ., 1983)

[**Answer.** The weights of oxygen are 0·08 gm and 0·16 gm respectively].

2. Name and state the Law of Chemical Combination which the following data illustrates :

Water and an oxide of sulphur contain 88·89% and 50% of oxygen respectively. Hydrogen sulphide contains 5·88% of hydrogen. *(Guru Nanak Dev Pre-Univ., 1983)*

[**Answer.** Law of Reciprocal Proportions].

3. If a certain oxide of nitrogen weighing 11 gm yields 5·6 litres of Nitrogen and another oxide of it weighing 15 gm gives the same volume of nitrogen (all measurements being taken at N.T.P. Show that this data illustrates the law of multiple proportions. *(Kurkshetra Pre-Univ., 1983)*

Chapter 4. Mole Concept

4. Arrange the following in the decreasing order of their weights in gms :

(i) $6·023 \times 10^{23}$ number of molecules of Hydrogen.

(ii) 3·0 moles of Neon.

(iii) 0·75 g-atom of Carbon.

Appendix 6 257

(*iv*) 11·20 litres of CO_2 at 0°C and 1 atmospheric pressure.

(*v*) 1 mole of oxygen.

(*Guru Nanak Dev Pre-Univ., 1983*)

[**Answer**. (*ii*), (*iv*), (*iii*), (*i*) and (*v*)]

5. Three containers of one litre capacity are filled as follows :

(*i*) 1·0 gm of hydrogen at 273°C.

(*ii*) 24·0 gm of oxygen at 0°C.

(*iii*) 16·0 gm of methane at −23°C.

Which container is at the greatest pressure ?

(*Guru Nanak Dev Pre-Univ., 1983*)

[**Answer**. The container with hydrogen is at the greatest pressure].

6. Determine the weight of :

(*i*) One silver atom.

(*ii*) 0·5 mole of H_2SO_4.

(*iii*) 1·0 litre of CO at N.T.P.

(*iv*) 2·0 mole of H_2O. (*Punjab Pre-Univ., 1983*)

[**Answer**. (*i*) $17·93 \times 10^{-23}$ gm ; (*ii*) 49·00 gm ; (*iii*) 1·25 gm; (*iv*) 36 gm.]

Chapter 8. Molecular Weights

7. 0·2 gm of a volatile substance displaced 30 cc of air at 27°C and 760 mm of pressure. Calculate the vapour density of the substance. (*Kurukshetra Pre-Univ., 1983*)

[**Answer**. 82·05]

Chapter 9. Equivalent Weights

8. The chloride of an element contains 49·50% of chlorine. The specific heat of the element is 0·059. Calculate the equivalent weight, valency, and atomic weight of the element.

(*Punjab Pre-Univ., 1983*)

[**Answer**. Eq. wt.=36·2 ; Valency=3; At. wt.=108·4]

9. For how much time a current of 2 amperes strength should be passed through a solution of copper sulphate to

deposite 0·636 gm of copper ?

(*Guru Nanak Dev Pre-Univ., 1983*)

[**Answer.** 16·23 minutes]

10. Find the equivalent weight of anhydrous $FeSO_4$ using the equation given below :

$$2\ FeSO_4 + H_2SO_4 + O \longrightarrow Fe_2(SO_4)_3 + H_2O$$

(Fe=56, S=32, O=16, H=1)

(*Kurukshetra Pre-Univ., 1983*)

[**Answer.** 152]

Chapter 10. Atomic Weights

11. Find the exact atomic weight of an element having specific heat 0·22 and equivalent weight as 9·20.

(*Kurukshetra Pre-Univ., 1983*)

[**Answer.** 27·60)

12. One atom of an element weighs $6·65 \times 10^{-24}$ gm. Name the element. (*Guru Nank Dev Pre-Univ., 1983*)

[**Answer.** Helium]

Chapter 11. Oxidation and Reduction

13. (*a*) What happens to the oxidation number of sulphur in the reaction :

$$H_2S + Br_2 \longrightarrow 2\ HBr + S$$

What type of chemical reaction is this ?

(*b*) Calculate the oxidation number of metals in the following :

(*i*) $Cr_2O_7^{2-}$ (*ii*) MnO_4^{2-} (*iii*) $[Fe(CN)_6]^{4-}$.

(*c*) Balance the following equations by O.N. method :

$$H_2S + O_2 \longrightarrow H_2O + SO_2$$
$$P + HNO_3 \longrightarrow HPO_3 + NO + H_2O$$

(*Guru Nanak Dev Pre-Univ., 1983*)

[**Answer.** (*a*) Oxidation number of sulphur changes from -2 to 0 (Oxidation of sulphur). This is an oxidation reaction ;

(b) (i) +3, (ii) +6, (iii) +2 ;
(c) $2 H_2S + 3O_2 \longrightarrow 2 H_2O + 2 SO_2$]

14. Balance the following equation by oxidation number method :

(i) $MnO_4^- + H^+ + Fe^{2+} \longrightarrow Mn^{2+} + Fe^{3+} + H_2O$

(ii) $Zn + NO_3^- + H^+ \longrightarrow Zn^{2+} + NH_4^+ + H_2O$

(*Punjab Pre-Univ., 1983*)

[**Answer**. (i) $MnO_4^- + 8 H^+ + 5 Fe^{2+} \longrightarrow Mn^{2+} + 5 Fe^{3+} + 4H_2O$

(ii) $Zn + NO_3^- + 10 H^+ \longrightarrow Zn^{2+} + NH_4^+ + 3H_2O$]

15. Calculate the oxidation number of Fe in Fe_3O_4 and $[Fe(CN)_6]^{3-}$. (*Kurukshetra Pre-Univ., 1983*)

[**Answer**. (i) 8/3 ; (ii) 3]

Chapter 12. Volumetric Analysis

16. 20·0 ml of 18 M H_2SO_4 diluted to a total volume of one litre. Find the molar concentration of this solution.

(*Punjab Pre-Univ., 1983*)

[**Answer**. 0·36 M]

17. What will be the weight of potassium hydroxide dissolved in 200 cc of 0·5 N solution of it. (K=39, O=16, H=1).

(*Kurukshetra Pre-Univ., 1983*)

[**Answer**. 5·6 gm]

Chapter 13. Problems Based on Equations

18. How many grams of $CaCO_3$ must be heated to get enough CO_2 which can convert 0·1 Mole of Na_2CO_3 into $NaHCO_3$. (Na=23, Ca=40, C=12, O=16, H=1).

(*Kurukshetra Pre-Univ., 1983*)

[**Answer**. 10 gm]

19. Calculate the mass of potassium chlorate which must be decomposed to give 224·0 mls of oxygen at N.T.P. (K=39, Cl=35·5, O=16). (*Guru Nank Dev Pre-Univ., 1983*)

[**Answer**. 0·81 gm]

Chapter 14. Problems Based on Equations

20. Find out the volume of Cl_2 at 27°C and one atmosperic pressure produced by the action of 200 ml of N/5 HCl on excess of manganese dioxide. (Mn=55, O=16, H=1, Cl=35·5).

(*Punjab Pre-Univ., 1983*)

[**Answer.** 246·15 ml]

21. One litre of oxygen at N.T.P. is made to react with three litres of CO N.T.P. Calculate the weight of each substance found after the reaction. (C=12, O=16).

(*Kurukshetra Pre-Univ., 1983*)

[**Answer.** $CO=1·25$ gm ; $CO_2=3·92$ gm]

LOGARITHMS

	0	1	2	3	4	5	6	7	8	9	1 2 3	4 5 6	7 8 9
10	0000	0043	0086	0128	0170	0212	0253	0294	0334	0374	5 9 13 4 8 12	17 21 26 16 20 24	30 34 38 23 32 36
11	0414	0453	0492	0531	0569	0607	0645	0682	0719	0755	4 8 12 4 7 11	16 20 23 15 18 22	27 31 35 26 29 33
12	0792	0828	0864	0899	0934	0969	1004	1038	1072	1106	3 7 11 3 7 10	14 18 21 14 17 20	25 28 32 24 27 31
13	1139	1173	1206	1239	1271	1303	1335	1367	1399	1430	3 6 10 3 7 10	13 16 19 13 16 19	23 26 29 22 25 29
14	1461	1492	1523	1553	1584	1614	1644	1673	1703	1732	3 6 9 3 6 9	12 15 19 12 14 17	22 25 28 20 23 26
15	1761	1790	1818	1847	1875	1903	1931	1959	1987	2014	3 6 9 3 6 8	11 14 17 11 14 17	20 23 26 19 22 25
16	2041	2068	2095	2122	2148	2175	2201	2227	2253	2279	3 6 8 3 5 8	11 14 16 10 13 16	19 22 24 18 21 23
17	2304	2330	2355	2380	2405	2430	2455	2480	2504	2529	3 5 8 3 5 8	10 13 15 10 12 15	18 20 23 17 20 22
18	2553	2577	2601	2625	2648	2672	2695	2718	2742	2765	2 5 7 2 4 7	9 12 14 9 11 14	17 19 21 16 18 21
19	2788	2810	2833	2856	2878	2900	2923	2945	2967	2989	2 4 7 2 4 6	9 11 13 8 11 13	16 18 20 15 17 19
20	3010	3032	3054	3075	3096	3118	3139	3160	3181	3201	2 4 6	8 11 13	15 17 19
21	3222	3243	3263	3284	3304	3324	3345	3365	3385	3404	2 4 6	8 10 12	14 16 18
22	3424	3444	3464	3483	3502	3522	3541	3560	3579	3598	2 4 6	8 10 12	14 15 17
23	3617	3636	3655	3674	3692	3711	3729	3747	3766	3784	2 4 6	7 9 11	13 15 17
24	3802	3820	3838	3856	3874	3892	3909	3927	3945	3962	2 4 5	7 9 11	12 14 16
25	3979	3997	4014	4031	4048	4065	4082	4099	4116	4133	2 3 5	7 9 10	12 14 15
26	4150	4166	4183	4200	4216	4232	4249	4265	4281	4298	2 3 5	7 8 10	11 13 15
27	4314	4330	4346	4362	4378	4393	4409	4425	4440	4456	2 3 5	6 8 9	11 13 14
28	4472	4487	4502	4518	4533	4548	4564	4579	4594	4609	2 3 5	6 9 9	11 12 14
29	4624	4639	4654	4669	4683	4698	4713	4728	4742	4757	1 3 4	6 7 9	10 12 13
30	4771	4786	4800	4814	4829	4843	4857	4871	4886	4900	1 3 4	6 7 9	10 11 13
31	4914	4928	4942	4955	4969	4983	4997	5011	5024	5038	1 3 4	6 7 8	10 11 12
32	5051	5065	5079	5092	5105	5119	5132	5145	5159	5172	1 3 4	5 7 8	9 11 12
33	5185	5198	5211	5224	5237	5250	5263	5276	5289	5302	1 3 4	5 6 8	9 10 12
34	5315	5328	5340	5353	5366	5378	5391	5403	5416	5428	1 3 4	5 6 8	9 10 11
35	5441	5453	5465	5478	5490	5502	5514	5527	5539	5551	1 2 4	5 6 7	9 10 11
36	5563	5575	5587	5599	5611	5623	5635	5647	5658	5670	1 2 4	5 6 7	8 10 11
37	5682	5694	5705	5717	5729	5740	5752	5763	5775	5786	1 2 3	5 6 7	8 9 10
38	5798	5809	5821	5832	5843	5855	5866	5877	5888	5899	1 2 3	5 6 7	8 9 10
39	5911	5922	5933	5944	5955	5966	5977	5988	5999	6010	1 2 3	4 5 7	8 9 10
40	6021	6031	6042	6053	6064	6075	6085	6096	6107	6117	1 2 3	4 5 6	8 9 10
41	6128	6138	6149	6160	6170	6180	6191	6201	6212	6222	1 2 3	4 5 6	7 8 9
42	6232	6243	6253	6263	6274	6284	6294	6304	6314	6325	1 2 3	4 5 6	7 8 9
43	6335	6345	6355	6365	6375	6385	6395	6405	6415	6425	1 2 3	4 5 6	7 8 9
44	6435	6444	6454	6464	6474	6484	6493	6503	6513	6522	1 2 3	4 5 6	7 8 9
45	6532	6542	6551	6561	6571	6580	6590	6599	6609	6618	1 2 3	4 5 6	7 8 9
46	6628	6637	6646	6656	6665	6675	6684	6693	6702	6712	1 2 3	4 5 6	7 7 8
47	6721	6730	6739	6749	6758	6767	6776	6785	6794	6803	1 2 3	4 5 5	6 7 8
48	6812	6821	6830	6839	6848	6857	6866	6875	6884	6893	1 2 3	4 4 5	6 7 8
49	6902	6911	6920	6928	6937	6946	6955	6964	6972	6981	1 2 3	4 4 5	6 7 8

LOGARITHM

	0	1	2	3	4	5	6	7	8	9	1 2 3	4 5 6	7 8 9
50	6990	6998	7007	7016	7024	7033	7042	7050	7059	7067	1 2 3	3 4 5	6 7 8
51	7076	7084	7093	7101	7110	7118	7126	7135	7143	7152	1 2 3	3 4 5	6 7 8
52	7160	7168	7177	7185	7193	7202	7210	7218	7226	7235	1 2 2	3 4 5	6 7 7
53	7243	7251	7259	7267	7275	7284	7292	7300	7308	7316	1 2 2	3 4 5	6 6 7
54	7324	7332	7340	7348	7356	7364	7372	7380	7388	7396	1 2 2	3 4 5	6 6 7
55	7404	7412	7419	7427	7435	7443	7451	7459	7466	7474	1 2 2	3 4 5	5 6 7
56	7482	7490	7497	7505	7513	7520	7528	7536	7543	7551	1 2 2	3 4 5	5 6 7
57	7559	7566	7574	7582	7589	7597	7604	7612	7619	7627	1 2 2	3 4 5	5 6 7
58	7634	7642	7649	7657	7664	7672	7679	7686	7694	7701	1 1 2	3 4 4	5 6 7
59	7709	7716	7723	7731	7738	7745	7752	7760	7767	7774	1 1 2	3 4 4	5 6 7
60	7782	7789	7796	7803	7810	7818	7825	7832	7839	7846	1 1 2	3 4 4	5 6 6
61	7853	7860	7868	7875	7882	7889	7896	7903	7910	7917	1 1 2	3 4 4	5 6 6
62	7924	7931	7938	7945	7952	7959	7966	7973	7980	7987	1 1 2	3 3 4	5 6 6
63	7993	8000	8007	8014	8021	8028	8035	8041	8048	8055	1 1 2	3 3 4	5 5 6
64	8062	8069	8075	8082	8089	8096	8102	8109	8116	8122	1 1 2	3 3 4	5 5 6
65	8129	8136	8142	8149	8156	8162	8169	8176	8182	8189	1 1 2	3 3 4	5 5 6
66	8195	8202	8209	8215	8222	8228	8235	8241	8248	8254	1 1 2	3 3 4	5 5 6
67	8261	8267	8274	8280	8287	8293	8299	8306	8312	8319	1 1 2	3 3 4	5 5 6
68	8325	8331	8338	8344	8351	8357	8363	8370	8376	8382	1 1 2	3 3 4	4 5 6
69	8388	8395	8401	8407	8414	8420	8426	8432	8439	8445	1 1 2	2 3 4	4 5 6
70	8451	8457	8463	8470	8476	8482	8488	8494	8500	8506	1 1 2	2 3 4	4 5 6
71	8513	8519	8525	8531	8537	8543	8549	8555	8561	8567	1 1 2	2 3 4	4 5 5
72	8573	8579	8585	8591	8597	8603	8609	8615	8621	8627	1 1 2	2 3 4	4 5 5
73	8633	8639	8645	8651	8657	8663	8669	8675	8681	8686	1 1 2	2 3 4	4 5 5
74	8692	8698	8704	8710	8716	8722	8727	8733	8739	8745	1 1 2	2 3 4	4 5 5
75	8751	8756	8762	8768	8774	8779	8785	8791	8797	8802	1 1 2	2 3 3	4 5 5
76	8808	8814	8820	8825	8831	8837	8842	8848	8854	8859	1 1 2	2 3 3	4 5 5
77	8865	8871	8876	8882	8887	8893	8899	8904	8910	8915	1 1 2	2 3 3	4 4 5
78	8921	8927	8932	8938	8943	8949	8954	8960	8965	8971	1 1 2	2 3 3	4 4 5
79	8976	8982	8987	8993	8998	9004	9009	9015	9020	9025	1 1 2	2 3 3	4 4 5
80	9031	9036	9042	9047	9053	9058	9063	9069	9074	9079	1 1 2	2 3 3	4 4 5
81	9085	9090	9096	9101	9106	9112	9117	9122	9128	9133	1 1 2	2 3 3	4 4 5
82	9138	9143	9149	9154	9159	9165	9170	9175	9180	9186	1 1 2	2 3 3	4 4 5
83	9191	9196	9201	9206	9212	9217	9222	9227	9232	9238	1 1 2	2 3 3	4 4 5
84	9243	9248	9253	9258	9263	9269	9274	9279	9284	9289	1 1 2	2 3 3	4 4 5
85	9294	9299	9304	9309	9315	9320	9325	9330	9335	9340	1 1 2	2 3 3	4 4 5
86	9345	9350	9355	9360	9365	9370	9375	9380	9385	9390	1 1 2	2 3 3	4 4 5
87	9395	9400	9405	9410	9415	9420	9425	9430	9435	9440	0 1 1	2 2 3	3 4 4
88	9445	9450	9455	9460	9465	9469	9474	9479	9484	9489	0 1 1	2 2 3	3 4 4
89	9494	9499	9504	9509	9513	9518	9523	9528	9533	9538	0 1 1	2 2 3	3 4 4
90	9542	9547	9552	9557	9562	9566	9571	9576	9581	9586	0 1 1	2 2 3	3 4 4
91	9590	9595	9600	9605	9609	9614	9619	9624	9628	9633	0 1 1	2 2 3	3 4 4
92	9638	9643	9647	9652	9657	9661	9666	9671	9675	9680	0 1 1	2 2 3	3 4 4
93	9685	9689	9694	9699	9703	9708	9713	9717	9722	9727	0 1 1	2 2 3	3 4 4
94	9731	9736	9741	9745	9750	9754	9759	9763	9768	9773	0 1 1	2 2 3	3 4 4
95	9777	9782	9786	9791	9795	9800	9805	9809	9814	9818	0 1 1	2 2 3	3 4 4
96	9823	9827	9832	9836	9841	9845	9850	9854	9859	9863	0 1 1	2 2 3	3 4 4
97	9868	9872	9877	9881	9886	9890	9894	9899	9903	9908	0 1 1	2 2 3	3 4 4
98	9912	9917	9921	9926	9930	9934	9939	9943	9948	9952	0 1 1	2 2 3	3 4 4
99	9956	9961	9965	9969	9974	9978	9983	9987	9991	9996	0 1 1	2 2 3	3 3 4

Log Tables

ANTILOGARITHMS

	0	1	2	3	4	5	6	7	8	9	1 2 3	4 5 6	7 8 9
·00	1000	1002	1005	1007	1009	1012	1014	1016	1019	1021	0 0 1	1 1 1	2 2 2
·01	1023	1026	1028	1030	1033	1035	1038	1040	1042	1045	0 0 1	1 1 1	2 2 2
·02	1047	1050	1052	1054	1057	1059	1062	1064	1067	1069	0 0 1	1 1 1	2 2 2
·03	1072	1074	1076	1079	1081	1084	1086	1089	1091	1094	0 0 1	1 1 1	2 2 2
·04	1096	1099	1102	1104	1107	1109	1112	1114	1117	1119	0 1 1	1 1 2	2 2 2
·05	1122	1125	1127	1130	1132	1135	1138	1140	1143	1146	0 1 1	1 1 2	2 2 2
·06	1148	1151	1153	1156	1159	1161	1164	1167	1169	1172	0 1 1	1 1 2	2 2 2
·07	1175	1178	1180	1183	1186	1189	1191	1194	1197	1199	0 1 1	1 1 2	2 2 2
·08	1202	1205	1208	1211	1213	1216	1219	1222	1225	1227	0 1 1	1 1 2	2 2 3
·09	1230	1233	1236	1239	1242	1245	1247	1250	1253	1256	0 1 1	1 1 2	2 2 3
·10	1259	1262	1265	1268	1271	1274	1276	1279	1282	1285	0 1 1	1 1 2	2 2 3
·11	1288	1291	1294	1297	1300	1303	1306	1309	1312	1315	0 1 1	1 2 2	2 2 3
·12	1318	1321	1324	1327	1330	1334	1337	1340	1343	1346	0 1 1	1 2 2	2 2 3
·13	1349	1352	1355	1358	1361	1365	1368	1371	1374	1377	0 1 1	1 2 2	2 3 3
·14	1380	1384	1387	1390	1393	1396	1400	1403	1406	1409	0 1 1	1 2 2	2 3 3
·15	1413	1416	1419	1422	1426	1429	1432	1435	1439	1442	0 1 1	1 2 2	2 3 3
·16	1445	1449	1452	1455	1459	1462	1466	1469	1472	1476	0 1 1	1 2 2	2 3 3
·17	1479	1483	1486	1489	1493	1496	1500	1503	1507	1510	0 1 1	1 2 2	2 3 3
·18	1514	1517	1521	1524	1528	1531	1535	1538	1542	1545	0 1 1	1 2 2	2 3 3
·19	1549	1552	1556	1560	1563	1567	1570	1574	1578	1581	0 1 1	1 2 2	3 1 3
·20	1585	1589	1592	1596	1600	1603	1607	1611	1614	1618	0 1 1	1 2 2	3 3 3
·21	1622	1626	1629	1633	1637	1641	1644	1648	1652	1656	0 1 1	2 2 2	3 3 3
·22	1660	1663	1667	1671	1675	1679	1683	1687	1690	1694	0 1 1	2 2 2	3 3 3
·23	1698	1702	1706	1710	1714	1718	1722	1726	1730	1734	0 1 1	2 2 2	3 3 4
·24	1738	1742	1746	1750	1754	1758	1762	1766	1770	1774	0 1 1	2 2 2	3 3 4
·25	1778	1782	1786	1791	1795	1799	1803	1807	1811	1816	0 1 1	2 2 2	3 3 4
·26	1820	1824	1828	1832	1837	1841	1845	1849	1854	1858	0 1 1	2 2 3	3 3 4
·27	1862	1866	1871	1875	1879	1884	1888	1892	1897	1901	0 1 1	2 2 3	3 3 4
·28	1905	1910	1914	1919	1923	1928	1932	1936	1941	1945	0 1 1	2 2 3	3 4 4
·29	1950	1954	1959	1963	1968	1972	1977	1982	1986	1991	0 1 1	2 2 3	3 4 4
·30	1995	2000	2004	2009	2014	2018	2023	2028	2032	2037	0 1 1	2 2 3	3 4 4
·31	2042	2046	2051	2056	2061	2065	2070	2075	2080	2084	0 1 1	2 2 3	3 4 4
·32	2089	2094	2099	2104	2109	2113	2118	2123	2128	2133	0 1 1	2 2 3	3 4 4
·33	2138	2143	2148	2153	2158	2163	2168	2173	2178	2183	0 1 1	2 2 3	3 4 4
·34	2188	2193	2198	2203	2208	2213	2218	2223	2228	2234	1 1 2	2 3 3	4 4 5
·35	2239	2244	2249	2254	2259	2265	2270	2275	2280	2286	1 1 2	2 3 3	4 4 5
·36	2291	2296	2301	2307	2312	2317	2323	2328	2333	2339	1 1 2	2 3 3	4 4 5
·37	2344	2350	2355	2360	2366	2371	2377	2382	2388	2393	1 1 2	2 3 3	4 4 5
·38	2399	2404	2410	2415	2421	2427	2432	2438	2443	2449	1 1 2	2 3 3	4 4 5
·39	2455	2460	2466	2472	2477	2483	2489	2495	2500	2506	1 1 2	2 3 3	4 5 5
·40	2512	2518	2523	2529	2535	2541	2547	2553	2559	2564	1 1 2	2 3 4	4 5 5
·41	2570	2576	2582	2588	2594	2600	2606	2612	2618	2624	1 1 2	2 3 4	4 5 5
·42	2630	2636	2642	2649	2655	2661	2667	2673	2679	2685	1 1 2	2 3 4	4 5 6
·43	2692	2698	2704	2710	2716	2723	2729	2735	2742	2748	1 1 2	3 3 4	4 5 6
·44	2754	2761	2767	2773	2780	2786	2793	2799	2805	2812	1 1 2	3 3 4	4 5 6
·45	2818	2825	2831	2838	2844	2851	2858	2864	2871	2877	1 1 2	3 3 4	5 5 6
·46	2884	2891	2897	2904	2911	2917	2924	2931	2938	2944	1 1 2	3 3 4	5 5 6
·47	2951	2958	2965	2972	2979	2985	2992	2999	3006	3013	1 1 2	3 3 4	5 5 6
·48	3020	3027	3034	3041	3048	3055	3062	3069	3076	3083	1 1 2	3 4 4	5 6 6
·49	3090	3097	3105	3112	3119	3126	3133	3141	3148	3155	1 1 2	3 4 4	5 6 6

ANTILOGARITHMS

	0	1	2	3	4	5	6	7	8	9	1	2	3	4	5	6	7	8	9
·50	3162	3170	3177	3184	3192	3199	3206	3214	3221	3228	1	1	2	3	4	4	5	6	7
·51	3236	3243	3251	3258	3266	3273	3281	3289	3296	3304	1	2	2	3	4	5	5	6	7
·52	3311	3319	3327	3334	3342	3350	3357	3365	3373	3381	1	2	2	3	4	5	5	6	7
·53	3388	3396	3404	3412	3420	3428	3436	3443	3451	3459	1	2	2	3	4	5	6	6	7
·54	3467	3475	3483	3491	3499	3508	3516	3524	3532	3540	1	2	2	3	4	5	6	6	7
·55	3548	3556	3565	3573	3581	3589	3597	3606	3614	3622	1	2	2	3	4	5	6	7	7
·56	3631	3639	3648	3656	3664	3673	3681	3690	3698	3707	1	2	3	3	4	5	6	7	8
·57	3715	3724	3733	3741	3750	3758	3767	3776	3784	3793	1	2	3	3	4	5	6	7	8
·58	3802	3811	3819	3828	3837	3846	3855	3864	3873	3882	1	2	3	4	4	5	6	7	8
·59	3890	3899	3908	3917	3926	3936	3945	3954	3963	3972	1	2	3	4	5	5	6	7	8
·60	3981	3990	3999	4009	4018	4027	4036	4046	4055	4064	1	2	3	4	5	6	6	7	8
·61	4074	4083	4093	4102	4111	4121	4130	4140	4150	4159	1	2	3	4	5	6	7	8	9
·62	4169	4178	4188	4198	4207	4217	4227	4236	4246	4256	1	2	3	4	5	6	7	8	9
·63	4266	4276	4285	4295	4305	4315	4325	4335	4345	4355	1	2	3	4	5	6	7	8	9
·64	4365	4375	4385	4395	4406	4416	4426	4436	4446	4457	1	2	3	4	5	6	7	8	9
·65	4467	4477	4487	4498	4508	4519	4529	4539	4550	4560	1	2	3	4	5	6	7	8	9
·66	4571	4581	4592	4603	4613	4624	4634	4645	4656	4667	1	2	3	4	5	6	7	9	10
·67	4677	4688	4699	4710	4721	4732	4742	4753	4764	4775	1	2	3	4	5	7	8	9	10
·68	4786	4797	4808	4819	4831	4842	4853	4864	4875	4887	1	2	3	4	6	7	8	9	10
·69	4898	4909	4920	4932	4943	4955	4966	4977	4989	5000	1	2	3	5	6	7	8	9	10
·70	5012	5023	5035	5047	5058	5070	5082	5093	5105	5117	1	2	4	5	6	7	8	9	11
·71	5129	5140	5152	5164	5176	5188	5200	5212	5224	5236	1	2	4	5	6	7	8	10	11
·72	5248	5260	5272	5284	5297	5309	5321	5333	5346	5358	1	2	4	5	6	7	9	10	11
·73	5370	5383	5395	5408	5420	5433	5445	5458	5470	5483	1	3	4	5	6	8	9	10	11
·74	5495	5508	5521	5534	5546	5559	5572	5585	5598	5610	1	3	4	5	6	8	9	10	12
·75	5623	5636	5649	5662	5675	5689	5702	5715	5728	5741	1	3	4	5	7	8	9	10	12
·76	5754	5768	5781	5794	5808	5821	5834	5848	5861	5875	1	3	4	5	7	8	9	11	12
·77	5888	5902	5916	5929	5943	5957	5970	5984	5998	6012	1	3	4	5	7	8	10	11	12
·78	6026	6039	6053	6067	6081	6095	6109	6124	6138	6152	1	3	4	6	7	8	10	11	13
·79	6166	6180	6194	6209	6223	6237	6252	6266	6281	6295	1	3	4	6	7	9	10	11	13
·80	6310	6324	6339	6353	6368	6383	6397	6412	6427	6442	1	3	4	6	7	9	10	12	13
·81	6457	6471	6486	6501	6516	6531	6546	6561	6577	6592	2	3	5	6	8	9	11	12	14
·82	6607	6622	6637	6653	6668	6683	6699	6714	6730	6745	2	3	5	6	8	9	11	12	14
·83	6761	6776	6792	6808	6823	6839	6855	6871	6887	6902	2	3	5	6	8	9	11	13	14
·84	6918	6934	6950	6966	6982	6998	7015	7031	7047	7063	2	3	5	6	8	10	11	13	15
·85	7079	7096	7112	7129	7145	7161	7178	7194	7211	7228	2	3	5	7	8	10	12	13	15
·86	7244	7261	7278	7295	7311	7328	7345	7362	7379	7396	2	3	5	7	8	10	12	13	15
·87	7413	7430	7447	7464	7482	7499	7516	7534	7551	7568	2	3	5	7	9	10	12	14	16
·88	7586	7603	7621	7638	7656	7674	7691	7709	7727	7745	2	4	5	7	9	11	12	14	16
·89	7762	7780	7798	7816	7834	7852	7870	7889	7907	7925	2	4	5	7	9	11	13	14	16
·90	7943	7962	7980	7998	8017	8035	8054	8072	8091	8110	2	4	6	7	9	11	13	15	17
·91	8128	8147	8166	8185	8204	8222	8241	8260	8279	8299	2	4	6	8	9	11	13	15	17
·92	8318	8337	8356	8375	8395	8414	8433	8453	8472	8492	2	4	6	8	10	12	14	15	17
·93	8511	8531	8551	8570	8590	8610	8630	8650	8670	8690	2	4	6	8	10	12	14	16	18
·94	8710	8730	8750	8770	8790	8810	8831	8851	8872	8892	2	4	6	8	10	12	14	16	18
·95	8913	8933	8954	8974	8995	9016	9036	9057	9078	9099	2	4	6	8	10	12	15	17	19
·96	9120	9141	9162	9183	9204	9226	9247	9268	9290	9311	2	4	6	8	11	13	15	17	19
·97	9333	9354	9376	9397	9419	9441	9462	9484	9506	9528	2	4	7	9	11	13	15	17	20
·98	9550	9572	9594	9616	9638	9661	9683	9705	9727	9750	2	4	7	9	11	13	16	18	20
·99	9772	9795	9817	9840	9863	9886	9908	9931	9954	9977	2	5	7	9	11	14	16	18	20